A STOVALL MUSEUM PUBLICATION

WILDERNESS BONANZA

The Tri-State District of
Missouri, Kansas, and Oklahoma

WILDERNESS BONANZA

The Tri-State District of Missouri, Kansas, and Oklahoma

by Arrell M. Gibson

University of Oklahoma Press : Norman

Wilderness Bonanza: The Tri-State District of Missouri, Kansas, and Oklahoma is published by the University of Oklahoma Press for the Stovall Museum, University of Oklahoma, Norman. Other Stovall Museum Publications are:

I *Oklahoma Archaeology: An Annotated Bibliography*, by Robert E. Bell, available in cloth and paper.

II *Reptiles of Oklahoma*, by Robert G. Webb, available in cloth and paper.

By Arrell M. Gibson

Published by the University of Oklahoma Press
The Kickapoos: Lords of the Middle Border (1963)
The Life and Death of Colonel Albert Jennings Fountain (1965)
Fort Smith: Little Gibraltar on the Arkansas (with Edwin C. Bearss) (1969)
The Chickasaws (1971)
Wilderness Bonanza: The Tri-State District of Missouri, Kansas, and Oklahoma (1972)

Library of Congress Cataloging in Publication Data

Gibson, Arrell Morgan.
Wilderness bonanza.

(A Stovall Museum publication)
1. Lead mines and mining—Great Plains. 2. Zinc mines and mining—Great Plains. I. Title. II. Title: The tri-state district of Missouri, Kansas, and Oklahoma. III. Series: Oklahoma. University. J. Willis Stovall Museum. A Stovall Museum publication.
TN453.A5G5 338.2'7'44 77-177335
ISBN 0-8061-0990-4
ISBN 0-8061-1033-3 (pbk.)

 Composed and printed at Norman, Oklahoma, U.S.A., by the University of Oklahoma Press. First edition.

To the Memory
of
Arrell Morgan (Pat) Gibson

Preface

The centennial observance in 1972 of Joplin, Missouri, is timely. It commemorates the first century of life, growth, and progress for the principal city of the Tri-State District. It also marks for Joplin and the region the close of one economic era and the beginning of another. The bonanza-rich mineral lodes for which Joplin and the Tri-State District became world famous are exhausted. Only marginal ores remain. Mining has ceased. Joplin in the 1970's enters a new economic age with new diversified industry.

A city's centennial observance necessarily considers its past and contemplates its future. Joplin and the Tri-State District have a unique heritage. Although the mineral bonanza has ended, the mining legacy, the values of pioneer miners, survives to give guidance for the new economic age. Thus it is timely to present the story of the legacy—natural and human—of Joplin and the Tri-State District.

Countless ages ago, two minerals of strategic use for a future civilization, lead and zinc, accumulated in great quantities over an area encompassing Jasper and Newton counties in southwestern Missouri, Cherokee County in southeastern Kansas, and Ottawa County in northeastern Oklahoma. As an economic region, it came to be called the Tri-State District. Its urban focus was Joplin, Missouri.

The mineral wealth of this region was first tapped commercially around 1850. Within one hundred years, Tri-State miners had extracted, processed, and sent into the markets of the world most of the readily accessible lead and zinc ores which nature had required several geologic ages to accumulate. The value of Tri-State mineral production from 1850 to 1950 exceeded one billion dollars, and until 1945 the region was rated as the leading producer of lead and zinc concentrates in the world, accounting for one-half of the zinc and one-tenth of the lead produced in the United States.

This century of regional mining development provides the substance for the rich heritage of Joplin and the Tri-State District. It is a unique heritage, too, because the history of Joplin and its satellite region—the Tri-State District—is unlike the history of any other mining region in the United States.

Wilderness Bonanza is possible because of the interest and assistance of many kind and gracious persons. Very special appreciation is expressed to the library and research staffs at the John Crerar Memorial Library, Chicago; the U.S. Department of Labor Library, Washington, D.C.; the U.S. Bureau of Mines Library, Washington, D.C.; the University of Missouri Library, Rolla, Missouri; the Oklahoma Historical Society Library, Oklahoma City; and the Bizzell Library, University of Oklahoma, Norman. I am particularly indebted to Leo Nigh, H. A. Andrews, Clarence Brown, Lester Davis, Roy Moore, Gene Hollon, Gilbert Fite, William F. Netzebrand, Jim Mueller, E. O. Humphrey, Dorothea Hoover, Jack Haley, and Joe Snell for special information, criticism, and direction. John Keever Greer, director of the Stovall Museum in the University of Oklahoma, and Duane Roller, chairman of the University of Oklahoma Faculty Research Committee, provided sustained encouragement and support. And I would be remiss if I did not acknowledge the superb work of my typist, Maryon Kaplan.

ARRELL M. GIBSON

Norman, Oklahoma
April 20, 1972

Contents

	Preface	*page* *ix*
I	The Region and Its Mineral Resources	3
II	District Development to 1865	14
III	District Development since 1865	27
IV	Tri-State Prospecting	41
V	Tri-State Mining Methods: The Early Period	67
VI	Tri-State Mining Methods: The Intermediate Period	79
VII	Tri-State Mining Methods: The Modern Period	91
VIII	Tri-State Milling	100
IX	Tri-State Smelting	113
X	Tri-State Land Tenure and Use Systems	127
XI	Tri-State Mine Finance	160
XII	Tri-State District Social Conditions	179
XIII	Tri-State Labor Relations: The Early Period	196
XIV	Tri-State Labor Relations: The Recent Period	210
XV	Tri-State Society	249
XVI	The Tri-State District at Mid-Century	266
	Notes	274
	Selected Bibliography	313
	Index	351

Illustrations

Lone Elm Mining Camp, Missouri *page* 51
Blende City Mining Camp, Missouri 52
Turkey Creek Mining Camp, Missouri 53
Carterville Mining Camp, Missouri 54
King William Mine, Duenweg Mining Camp, Missouri 55
Pure galena from the Bradbury Mine 56
Opening an ore vein, Netta Mine 57
Primitive horse hoister, Tanyard Hollow Camp 58
Commerce Mining Camp, Oklahoma 60
Churn-drill prospecting, Oklahoma 62
Churn-drill prospecting, Missouri 63
Webb City Mining Camp, Missouri 64
Joplin Mining Camp, Missouri 64
Eagle Picher Mining and Smelting Company, Empire City 65
Badger Mine, Empire City 131
Panoramic scenes in Picher Mining Camp, Oklahoma 132
Picher Mining Camp, the last boom town 134
Netta Mine, Picher-Cardin Mining Camp, Oklahoma 136
Daisy D. Mine, Empire City 138
Morning Star Mine, Empire City 139
Empire City, Galena Mining Camp, Kansas 140
Empire City, Galena Mining Camp, Kansas 142
Empire City, Galena Mining Camp, Kansas 144
Treece Mining Camp, Kansas 146
Webb City Mining Camp, Missouri 211
Picher Mining Camp, Oklahoma 212
Chitwood Mining Camp, Missouri 213

Open-pit mining 214
Leadville Hollow Camp, Missouri 215
Duenweg Mining Camp, Missouri 216
Opening a mine in Missouri 217
Shoveler crew 218
Chitwood Mining Camp, Missouri 219
Oronogo Mining Camp, Missouri 220
Stotts City Mining Camp, Missouri 222
Prosperity Mining Camp, Missouri 223
Rich Hill Mine, Missouri 224
Swindle Hill Mine, Missouri 226

MAPS

Tri-State Mineralized Area *page* 10
Country of the Six Boils, 1836 18
Tri-State District, 1860 22
Principal Towns in the Tri-State District, 1950 268

WILDERNESS BONANZA

The Tri-State District of Missouri, Kansas, and Oklahoma

CHAPTER I

The Region and Its Mineral Resources

The Tri-State Lead and Zinc Mining District extends over an area of approximately 1,188 square miles—from Wentworth, Missouri, in the east, to the Neosho River between the Melrose Camp and Chetopa, Kansas, on the west; to the north from Neck City, Missouri, on the North Fork of Spring River, southward to Newtonia, Missouri; and southwesterly to Miami, Oklahoma. The district embraces Jasper and Newton counties in Missouri; Cherokee County, Kansas; and Ottawa County, Oklahoma. The Tri-State Mining District is situated on the northwest side of the Ozark Uplift, dropping gently into the Spring River Valley, and from there rising briefly westward over the Neosho River divide in Kansas and Oklahoma. This area's surface elevation declines from about 1,200 feet above sea level on the east to 800 feet on the west.[1] Streams in the district, containing clear, potable water, include Spring River, Neosho River, Elk River (sometimes called Cowskin), Five Mile Creek, Shoal Creek, Jenkins Creek, Cedar Creek, Jones Creek, Turkey Creek, Joplin Creek, and Center Creek. Spring River drains the greater portion of the region.[2]

The relief of the district is gently rolling prairie punctuated by brief hills giving purchase to the creek and river valleys. These promontories are heavily wooded with mixed stands of timber; white oak and blackjack predominate. The hill country affords considerable pasture where clearings have been made. In spots the prairie still holds its virgin sod and produces excellent hay crops. The prairies and the river valleys yield good grain and forage crops when cultivated. Climatically the region is classified as temperate continental. Its mildness permitted mining with few interruptions through the entire year.[3]

A variety of names other than Tri-State District have been used to designate this mining region of southwest Missouri, southeast Kansas, and northeast Oklahoma. From 1850 until 1876 it was called the Southwest District of Missouri, since the field of production was exclusively within the Missouri section. Following discoveries of ore in Kansas and Indian Territory (after 1907, Oklahoma), the region was designated the Missouri-Kansas, Joplin, and Missouri-Kansas-Oklahoma District, although Joplin District is still a common designation.[4]

Joplin District is not a misnomer, however, since all mining camps in the three-state area came to be subsidiary to Joplin, Missouri. Banking, merchandising, transportation facilities, the manufacturing and supply of mining equipment, and ore smelting and marketing facilities radiated from Joplin. This city housed most of the mining company offices, the research facilities, and a large portion of the district's labor force.[5]

There is no universally accepted explanation of the origin of the lead and zinc ores of the Tri-State District. How the ores were transported, concentrated, and enriched throughout the region in the course of geologic ages has been studied by at least fourteen eminent geologists. One view emerging from these studies is that the ores were concentrated into rich deposits by cold meteoric waters which include descending ground waters and ascending artesian waters. Another view is that the Tri-State lodes were produced by ascending thermal waters of magmatic origin; that is, warm to hot waters carried molten rock material from within the earth's center to upper levels where the solution ultimately cooled.[6]

Regardless of the theory accepted, it must account for the reservoirs into which the minerals could be deposited. The cause and nature of these once-empty openings in which the minerals collected are not a source of controversy. Geologists agree that the Ozark Uplift was the scene in past geologic ages of considerable faulting, flexing, brecciation or shattering, and various other physical movements, each of which produced structural deformations in the strata of the district. These deformations included the opening of caves, cavities of various shapes and volume, channels, slips, and crevices.[7] Through these openings, water and other liquids could circulate.

The water deposition theory is somewhat as follows. Shallow seas over the region precipitated mineral materials through organic matter. Evaporation might have helped concentrate the solution.[8] These metallic materials saturated the limestone beds that today underlie the district. Ascending artesian waters, containing carbon dioxide and hydrogen sulfide, picked up the widely dispersed minerals and carried them in solution upward to faulted rock reservoirs. The lead and zinc crystallized out and cemented the shattered rock fragments. This process is called primary enrichment by H. F. Bain and C. R. Van Hise.[9]

Secondary concentration or enrichment was produced by the action of descending waters. Surface and ground waters were involved here. Each carried a small quantity of lead and zinc material suspended in solution. These flowed through the underground channels and collected in the faulted and flexed reservoirs. According to W. S. T. Smith, this secondary concentration is still taking place.[10] Clarence A. Wright contends that this vertical, then lateral, circulation of mineralized waters explains the general district location of lead and zinc. Zinc, more soluble than lead, in solution is carried downward and is found at a greater depth. It is probable that "where a large body of galena is found with scattered amounts of blende [zinc], there will be a good deposit of blende below it."[11]

Those who follow the water deposition theory offer as evidence the appearance of minerals in ground waters, since, on analysis, well and spring tests show some zinc content. Tests on waters from two springs near Shoal Creek, one of the region's major streams, disclosed 297.7 parts per million of zinc.[12] The research of Smith on district springs showed a variance of 12 to 133 parts for every million of zinc.[13] The sediment in reservoirs at the Pittsburg, Kansas, waterworks was examined by C. E. Siebenthal for mineral trace. He calculated that eight and one-quarter pounds of zinc had accumulated during a year's time within the reservoir basins. He added that this did not account for all the zinc present, since on analysis the water demonstrated a one-part-per-million zinc content, making a total of 1,600 pounds of zinc in the water pumped during the year.[14]

The magmatic theory of origin of the Tri-State ores, advocated by George M. Fowler, Joseph P. Lyden, William A. Tarr, and Samuel Weidman, is a more recent concept. Briefly it consists of the following. Deep within the earth's viscera a body of molten material called magma developed. The heat of this mass increased the temperatures of surrounding rocks, and the gases and vapors of the magma intruded into the pores of peripheral material. A combination of the magma's heat, gases, and vapors produced a chemical reaction on the mineral material in the surrounding rock, forcing a sort of concentration of ores, including lead and zinc. Heated artesian waters carried this molten, magmatic solution upward into the reservoirs of the Ozark Uplift, created by faulting and flexing, where a cooling process localized it.[15]

Lead and zinc were the two principal Tri-State minerals, but there were others of commercial importance. A comprehensive cataloging of local minerals, made by A. F. Rogers in 1903, listed and described thirty-nine different types resident there.[16] Some of the more important minerals, in addition to lead and zinc, are as follows. In the northern portion of the district, sandstone and rock asphalt are mined. South of Joplin, dolomite (magnesium carbonate)—53 to 55 per cent calcium and 40 to 43 per cent magnesium—is quarried. North of Joplin, clays and shales are found in association with coal beds and are used in the manufacture of fire clay, bricks, and stoneware. Gravels and chats derived from the milling of lead and zinc ores, grouped like low-lying mountains around the mines, are widely used as railroad ballast and concrete material. From 1942 to 1950, shipments of mill gravel averaged 4,129,000 tons annually.[17]

Limestone, which is quarried for building and ornamental purposes, underlies the entire district. The so-called Carthage Stone takes a high polish and is seldom equalled for beauty. Near Seneca, Missouri, there are deposits of tripoli (silicon dioxide) that produce a soft, porous silica used in polishing metals, manufacturing emery, and filtering water. A pre–Civil War observer, Wiley Britton, reported that bituminous coal was mined in small quantities for blacksmithing purposes.[18]

In addition to thin coal seams, there are bitumen deposits that produce tar springs, the most famous one occurring on Tar Creek in the mining camp of Cardin, Oklahoma.[19] Wright noted that the Joplin mines near Chitwood contained some bitumen deposits, occupying small openings and crevices, which were usually viscous at ordinary mine temperatures.[20] In the newly opened Oklahoma mines it was reported that bitumen produced concentration difficulties in that it "balled up" the ore in the milling apparatus and settled as a thin film over the ore. This condition lowered the grade of the concentrates and thus reduced the price per ton.[21]

Other minerals found in the area that caused mining-milling difficulty were marcasite and pyrite (sulfides of iron). When subjected to oxidation, these minerals decompose into melanterite in the mine workings and take on the form of stalactites and stalagmites on mine walls and roofs. Marcasite, when oxidized by aerated mine water, put out so much heat that operators had to install special ventilating equipment in the drifts.[22] These iron sulfides are nearly as heavy as zinc ore and made separation by jigging difficult. Also, they lowered the ore grade, as there was a penalty of one dollar a point per ton for iron content.[23]

Small quantities of quartz and jasperoid, a dark formation, are disseminated through the ore-bearing rock. Calcite and aragonite are forms of calcium carbonate found in the ore deposits and act as a cement for the ore in the cavities. Crystals of calcite often reach three to four feet in length in some of the large caves.[24] These and dolomite (carbonate of calcium and magnesium), locally called tiff, frequently caused trouble in milling because the specific gravity is higher than that of the flint and other gangue materials, and small amounts were found in the zinc concentrates.[25] Some fluorine has been discovered in the concentrates through smelting. One of the refiners of local ores, the National Zinc Company at Bartlesville, Oklahoma, had trouble in its acid plant because of the fluorine content in Tri-State-produced ores.[26]

Of course, the chief commercial minerals of the Tri-State District were lead and zinc. These minerals were found in several forms. For zinc these were: sphalerite, or zinc sulfide, locally known as jack, rosin jack, black jack, and blende; smithsonite, or zinc carbonate, locally designated zinc dry-bone and carbonate; calamine or zinc silicate, locally called silicate; and hydro-zincite, not found in commercial quantities. Calamine was of considerable importance in the early mining history of the Peoria and Lincolnville camps in Oklahoma. These nonsphalerite types were due to weathering of sphalerite above the ground-water level in the upper reaches of the mines.[27] Sphalerite was the most important mineral of the district. It contained about 63 per cent metallic zinc and 33 per cent sulfur. The yellowish streaks, manifest in most specimens of this material, indicate the high sulfur content.[28]

Lead was the other mineral produced commercially in the Tri-State District. Technically known as galena, it is lead sulfide, containing about 13 per cent sulfur and 87 per cent lead. The lead of this area was peculiarly pure and nonargentiferous, containing practically no silver.[29] In other lead-producing areas of the world this metal contains traces of silver, arsenic, antimony, and similar impurities. "It is probably true that no mining locality in the world produces a purer galena or one more easily smelted," Erasmus Haworth said of the Tri-State District in his *Report for 1897*.[30] Other forms of lead found in lesser quantities were lead carbonate, locally called dry-bone or carbonate; pyromorphite, or lead phosphate, sometimes designated green lead; and anglesite, or lead sulfate. These secondary ores were the result of weathering and were always found near the surface.[31]

The lead and zinc ores of the Tri-State District can be separated only superficially for purposes of description. In their natural state these minerals were intimately associated. Few paying shafts produced amounts of one without likewise producing some quantities of the other.[32] Every mining and milling process used in the district was adapted for processing both minerals in the same operation. In terms of vertical occurrence the lead ore generally was closer to the surface, the zinc ore deeper. For the most part, however, these ores were physically connected, and specimens examined disclosed lead and zinc associated in the same piece of chert.

A résumé of the geological character of the Tri-State District is essential to understanding some of the problems involved in prospecting, mining, and milling. At one time probably all the region was understructured with limestone.[33] Through geologic ages this primary material has been modified by the intrusion of chert of various colors, locally called flint, and minerals including lead and zinc. In the days of shallow mining, the belief was prevalent among miners that ore bodies rarely, if ever, occurred beneath a body of limestone. It was common to abandon a shaft if limestone appeared. According to Haworth, this was a wise policy if the operators lacked sufficient capital to carry the shaft through the limestone overburden, which sometimes reached a thick-

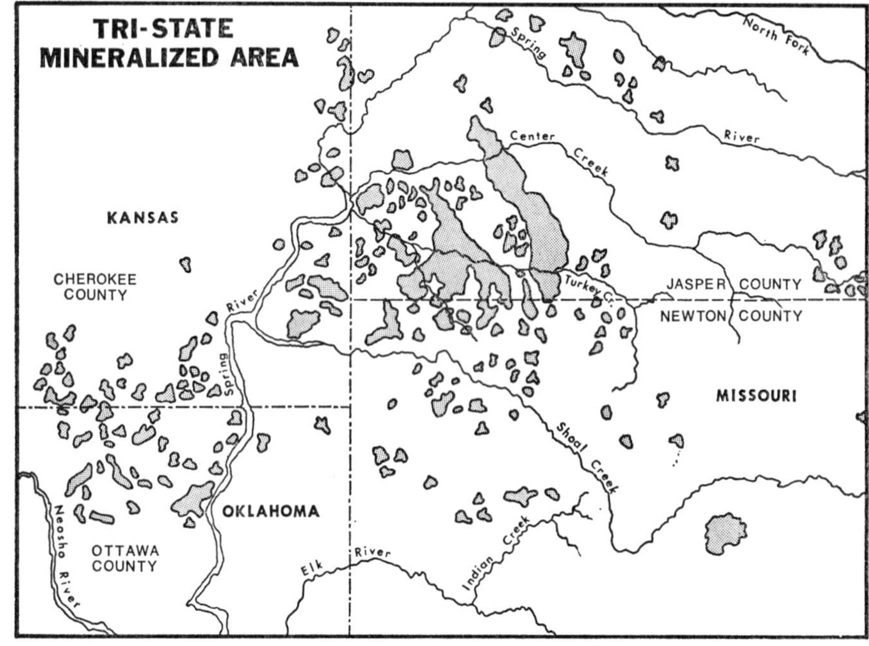

ness of fifty feet of solid material.[34] Drilling disclosed, though, that this limestone cap often covered rich lodes.

The Tri-State District contained a variety of ore deposits, each having peculiar qualities requiring a variety of exploitation methods. The shape and richness of deposits varied as did the ground surrounding them. Three different types of ground were discernible in the district.

(1) *The brecciated and boulder ground.* Locally known as hard ground, it consisted of massively bedded rock and predominated in the Kansas-Oklahoma section of the region. The lead and zinc minerals were mixed with flint in the fissures of these rocks.[35] The hardness of this material necessitated heavy blasting to loosen it after powder holes had been opened by machine drills.[36] The durability of this ground type allowed

mining underground with little timbering except periodic pillars to prevent roof caving. These pillars consisted of native rock left in the drift for roof support.

(2) *Sheet ground.* This ground type was chiefly found in the Webb City, Carterville, Oronogo, Porto Rico, and Duenweg areas, although its characteristics were evident in such camps as Baxter Springs, Kansas.[37] Sheet-ground qualities included a horizontal flint bed approximately twenty feet thick through which thin seams of ore circulated.[38] Sheet ground was notorious for its caving tendency; therefore timbering as well as frequent pillars were required to produce minimum safety underground.[39] One of the explanations for the hazards of mining in sheet ground was the prevalence of cotton rock,[40] a white form of decomposed flint manifest in the roofs of sheet-ground mines. Where it appeared, the cotton rock had to be removed as waste with the ore, or frequent roof supports were necessary.[41]

(3) *Soft ground.* The earliest mines operated in this type of formation, miners seeking the rich and readily accessible minerals close to the surface. A higher percentage of lead, as a general rule, was drawn from this type of deposit.[42] The depth of operations seldom exceeded 150 feet. The rock and soil constituents made this type so unstable that it was rather easily loosened with pick and bar, followed by a light powder charge.[43]

As late as 1950 small operators continued to exploit the soft-ground deposit with some profit, since its shallowness required only a small capital investment. Haworth reported in 1897 that, in the Short Creek Valley of the Galena camp, deeper levels revealed extensive ore deposits one hundred to two hundred feet below early soft-ground workings that had been abandoned when worked out.[44] This indicated that toward the surface soft-ground workings often overlay hard-ground deposits and that both types existed in the same general area, one paralleling the other.

Several different types of deposits based on shape were found in each kind of ground. There was, however, some degree of localization in the type of deposit. Geographically, for example, blanket deposits were found in the Webb City–Carterville–Porto Rico–Duenweg area, while the circle type of deposit was found in the vicinity of Alba and Oronogo. The flat and vertical run type predominated in the Oklahoma section of the field. The officially recognized types of ore deposit forms were (1) pocket formation, sometimes called circle deposit; (2) blanket veins; (3) flat runs; and (4) vertical runs.[45]

The pocket formation was characterized as capricious, since it constituted a disconnected concentration of ore apart from the main ore-bearing formation. It was frequently circular in shape. As a general rule the pocket vein was a small producer, although drill prospecting often gave the appearance of a heavier run. Of course, pocket formations could be heavy producers if the pockets appeared close enough together. Some single pocket or circle formations were exceedingly rich. Most notable of these was the Great Circle Mine at Oronogo.[46]

Blanket ore deposits predominated in sheet ground, although these appeared in other ground formations.[47] These veins were so called because they varied in thickness from that of a knife blade up to twenty feet[48] and "extend[ed] with remarkable persistence over wide areas," in blanket-like fashion. The most extensive blanket vein deposit in the world was developed in 1908, around and through Webb City. This famous ore bed extended in a continuous line for twelve miles. The chief disadvantage of the blanket deposit was its traditionally low-concentrate yield, which averaged from 2⅓ to 3 per cent of the ore mined, whereas nonblanket ore deposits in the district yielded as high as 10 to 50 per cent concentrates for the ore mined.[49]

Flat or horizontal runs varied with the geological conditions of structure and mineral concentration. As a general rule, clearly defined flat runs ranged from fifteen feet up to five hundred or one thousand feet in width and from ten to thirty feet in height. This kind of deposit might constitute a single layer, or, as was often the case, it might contain layers of minerals at various levels, generally parallel, and requiring drifting from the shaft at appropriate points.

The vertical run type of ore deposit had steep walls and was largely the result of ore concentrated in a rock fracture or fissure. Its dimensions varied in width from 10 to 15 feet, and often attained 150 feet in height.[50] The run type of ore deposit was found throughout the Tri-State District, although it predominated in the Miami section of the field and around Baxter Springs, Kansas.[51]

CHAPTER II

District Development to 1865

To understand the evolution of the Tri-State Mining District as the world's leading producer of lead and zinc concentrates, it is essential to gain a perspective from the history of the early development of eastern Missouri mines. When the southwestern district was opened around 1850, the geological skill, mining methods, and smelting techniques applied there were not produced by some act of legerdemain. Nor did these mining essentials emerge spontaneously from conditions inherent in the Southwestern environment. Rather, successful commercial mining was carried on as a result of the arrival of professional miners and settlers from other mining areas, especially eastern Missouri.

This famous eastern mining region was exploited for the first time by the French. In 1712, Louis XIV granted to the court favorite, Anthony Crozat, a commercial monopoly of an area embracing the present states of Louisiana, Mississippi, Tennessee, Missouri, Arkansas, and Illinois. This grant included exclusive rights to any minerals found there, provided the grantee developed mines within a period of three years. Crozat's patent retroceded to the crown, and in 1717 a group of Paris promoters, called the Company of the West and headed by the fabulous John Law, received much the same trading and mineral rights. The name Crozat Patent, however, continued to be applied to the mining portion of the grant.[1]

The Company of the West induced farmers, miners, and mechanics to immigrate to the new land. Among the early French emigrants was the son of a French miner, Philip

Renault. He headed the Company of St. Philip, an adjunct of the Company of the West, established for the purpose of conducting mining operations in the New World. Renault left France in 1719 with two hundred miners, tools, and supplies. At Santo Domingo he purchased five hundred slaves to assist his miners in working the anticipated deposits. Arriving at New Orleans in 1720, Renault and his party journeyed to the French outpost of Kaskaskia in Illinois. Using this settlement as a headquarters, he sent prospecting parties into French Louisiana.[2]

One of Renault's aides, M. La Motte, while exploring on the St. Francis River in eastern Missouri during 1720, discovered the famous lead mines that carry his name to this day. La Motte's strike produced a flurry of prospecting and discoveries throughout eastern Missouri. The metal was smelted on the spot by log furnaces and carried by pack horses to the Mississippi, floated on rafts to New Orleans, and exported to France.[3] While American miners invariably molded lead in pigs—molds of approximately five inches deep, five inches wide, and about sixteen inches long—the French and Spanish miners are reported to have molded the refined lead in rings. Walter Williams and Lloyd C. Shoemaker speculated that these rings of lead were made so they could be carried by slipping them on the necks of pack animals, one lead ring to an animal, thereby leaving the back free for the loading of additional merchandise.[4]

When Louisiana was transferred to Spain in 1762, Spanish and French miners continued operations on much the same basis as before in the eastern Missouri workings. Henry R. Schoolcraft found their methods exceedingly crude. Not more than 50 per cent of the metal was drawn from the ore. The wasteful open log furnace was the only extractor-type used, and valuable lead ash was thrown away as debris. A production report of 1773 reveals that St. Louis was the major Spanish lead depot in northern Louisiana. In that year 178 quintals (about 17,800 pounds) of lead were produced and delivered by miners in Missouri to St. Louis.[5]

A Virginia miner, Moses Austin, came to Missouri in 1797 to assist in the improvement of mining techniques. In return for sharing his skills in mining and erecting reverberatory or blast furnaces, Austin received one league square of eastern Missouri mining land from the Spanish government. Beginning operations in 1798, Austin set up furnaces, introduced a means for smelting the lead ashes, and sank the first conventional shaft with drifts for raising ore.[6]

Austin's innovations increased the output of refined lead in proportion to raw material. He found that the crude Franco-Spanish smelting methods yielded 250 pounds of refined lead to 1,000 pounds of mineral. Austin was able to derive 60 to 70 per cent of the lead ore smelted.[7]

Franco-Spanish tenure in Louisiana ended in 1803, when the United States purchased this resource-laden region. Thereupon, Americans flowed west developing farms, industries, and mines. The old French and Spanish mining fields were then exploited chiefly by these newcomers. At first, mining was concentrated in the old Crozat Patent area of eastern Missouri.

Edward Tiffin inspected the eastern Missouri mines in 1804, but he failed to mention exploration or mining operations in the Southwest. His description leaves the impression that the eastern field was the only known lead-producing area in the territory.[8] Yet Austin mentioned lead mines two hundred miles west of the Mine La Motte region of eastern Missouri. This would place the Austin reference in the present Tri-State District.[9] Tiffin found the eastern field so promising that he sent a communication to Congress in 1804, through Benjamin Stoddard, describing the richness of these mines and the prospect of completely meeting all national lead needs from them.[10]

Schoolcraft visited these eastern Missouri camps in 1818

and described in some detail the early American operations. He also journeyed to the future Tri-State District, noting that it was "a howling wilderness," the home of the wandering hunter and trapper. Here he found good lead specimens in the creek banks as well as traces of shallow mining and crude log furnaces presumably used by Indians and hunters. Lead was a major product of trade in the Southwest. Schoolcraft himself mined some lead on the banks of local streams, smelting the shallow-mined ore in a brief pit, insulating its sides with flat stones secured from the river beds. In 1818 he built a crude cabin on the fringe of the district, aware that, except for occasional trappers and Indians, he was the only human occupant within two hundred miles.[11]

Members of the Stephen H. Long expedition skirted the future Tri-State District in 1819–20. Their reports mentioned the geology of the Ozark area and noted traces of lead ores.[12] Schoolcraft returned from his second trip to the Southwest in 1819 and reported that he had mined lead, smelted it, and molded bullets right on the spot.[13] The fact that the mineral smelted and molded so readily indicates the innate purity of local ore.

It is certain that the "howling wilderness" of southwest Missouri had its mineral wealth tapped by trappers and traders near present Oronogo around 1838. John R. Holibaugh reported the discovery of mining traces in this vicinity where rains had eroded upper strata material. This surface lead, often exposed at the ground level, or as frequently found in the grass roots, was melted by wood-chip fires and molded into bullets. Some lead was smelted into pigs suitable for personal transportation and used as trading media.[14]

No permanent settlement took place in this region until 1833. Edmund Jennings, a hunter from Jackson County, Tennessee, is reported to have visited the region sometime during 1815. Returning home, Jennings described to his neighbors the natural beauty and resources of this wilderness

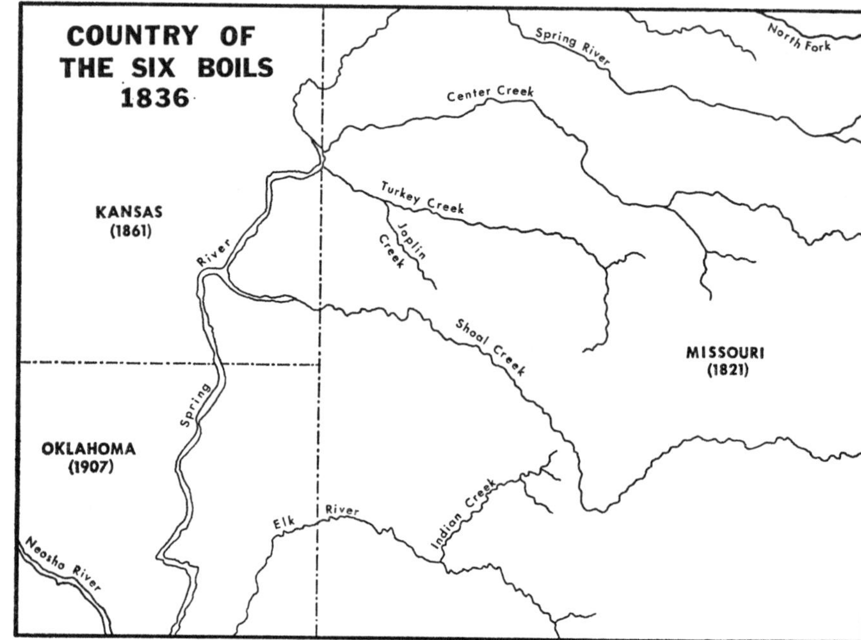

which he called "the land of the six boils" (from the boiling, surging qualities of the clear and swift Indian Creek, North Fork, Elk River, Center Creek, Turkey Creek, Shoal Creek, and Spring River). In Jennings' audience was a lad of fifteen, John C. Cox. Apparently enthralled by Jennings' account of southwestern Missouri, Cox journeyed to the region in 1838.[15]

As the first settler, Cox built a log dwelling within the present limits of Joplin and established a post office, calling it Blytheville.[16] According to Joel T. Livingston, Blytheville was named for a Cherokee Indian called Blythe who lived on nearby Shoal Creek.[17] Soon additional settlers were attracted to the Six Boils Country, and, in 1841, John Cox built a general store adjacent to the Blytheville post office to serve the locality. The settlement received the services of a pastor in 1839 when

Harris Joplin, a Methodist minister, arrived. He built a cabin and place of worship on the banks of a spring-fed run which the settlers soon named Joplin Creek.[18]

The initial settlers were primarily agriculturalists, attracted by this region's plentiful resources. Abundant wildlife furnished a dependable food supply. From the heavily timbered hillsides, material for shelter and fuel was readily available. The river valleys contained excellent cropland, and the prairie grasslands furnished superb pasture for livestock. Surface-exposed pieces of lead ore, the future source of wealth, were gathered and melted into shot by these early settlers.[19] While no strong hint of the exceedingly rich ore beneath their farms presented itself, as 1850 approached, more and more evidence of mineral wealth appeared.[20]

An example was the alleged experience of Levi Lee in 1844. This Newton County pioneer was hunting deer on the banks of Shoal Creek. As he stepped too close to the water's edge, the soft ground caved, and to avoid falling into the creek, he seized a clump of short bushes nearby. The weight of Lee's body is supposed to have pulled the clump out by the roots, yet his fall was broken. Recovering his balance, Lee noted that he had excavated several pieces of cubed material, which on closer examination were found to be lead ore. Further probing disclosed a complete ledge of ore-bearing rock a few feet down. The account continues that he did nothing to develop the discovery beyond a periodic visit to obtain lead for melting into shot.[21]

Another district discovery, the first to involve commercial mining, was reported by William Tingle in 1848.[22] The *Illustrated Historical Atlas of Jasper County* indicates that Tingle owned a farm near what became Leadville, two miles northwest of the present city of Joplin. Tingle was the first settler-farmer to engage actively in mining.[23]

On the heels of Tingle's discovery, professional and amateur prospectors set out in earnest to find new deposits. David

Campbell, a skilled lead miner from the Washington County field of eastern Missouri, visited John Cox in 1849 and pointed out that the latter's property contained favorable depressions and geologic indications of ore.[24] Almost simultaneous with Campbell's disclosure, lead was found on Cox's property. Dolph Shaner presented the following account of the Cox strike, based on an interview with John Cox's son. Cox owned a Negro youth who frequently fished in Joplin Creek. While digging for bait, in what later became the Kansas City Bottoms, the boy excavated some heavy cubed material which he carried to his master. Cox made a fire in an impromptu furnace and smelted the heavy material into refined lead.[25] This slave's discovery in Joplin Creek Valley gave rise to the label of "nigger diggins" as the field developed. The Tingle and Cox discoveries produced new interest in Missouri's southwest. Each week following the discoveries witnessed new settlers, now for the most part experienced miners. Additional discoveries were made around the Tingle and Cox mines, and a small mining camp called Leadville mushroomed with a population of one hundred persons in 1850.[26]

These professional miners were joined by local farmers. Moses Austin noted in 1804 that eastern Missouri farmers, when seasonally unemployed, literally dug in their door yards as much lead ore as was required to purchase important needs for their families.[27] Southwest Missouri farmers continued this practice. Experienced miners from the fields of Iowa, Illinois, and eastern Missouri joined local residents in the quest for minerals.[28] Some settlers, however, remained farmers and made a good living furnishing food and mine timbers to the mining population.[29]

As the field expanded, the Cox and Leadville mines came to be referred to in the *Reports of the Geological Survey of Missouri* as the Turkey Creek Mines of Jasper County. Soon other points in the county became centers of exploration and mining. Oronogo was one of the first, and conflicting accounts describe the discovery of mineral there. Holibaugh stated that mining and smelting were conducted near there on Center Creek in 1836 by trappers and Indians for the production of shot.[30] The *Historical Atlas* mentions that a mine and smelter were operated on Center Creek in 1849.[31] Joel T. Livingston and F. A. North wrote that Andrew McKee and Thomas Livingston found lead ore near the surface within the present limits of Oronogo in the spring of 1851 while excavating for a cellar. Livingston and North's exclusion of the earlier accounts indicates that they regard 1851 as the earliest discovery date.[32] Certainly the McKee-Livingston strike was the initial one within the present limits of Oronogo.

The mining camp that followed these discoveries took the name of Minersville in the pre–Civil War era, and its tributary mines were scattered over a radius of five miles.[33] This prosperous little community, which contained twenty-five dwellings in 1861, was desolated by the struggle that followed. Before its destruction, however, Minersville became the central shipping point for refined lead from the entire district. This camp claimed the largest chunk of pure lead found anywhere in the region. Excavated only six feet down, the huge piece of pure galena had to be broken up to be hoisted since it weighed thirty thousand pounds.[34]

Here, as well as in other camps, zinc ore was frequently encountered in conjunction with the lead. Little was known of this associate mineral's value in the district. Winslow observed huge quantities of it on debris piles around the mines in 1854.[35] Lack of interest in the zinc may be partly attributed to its limited commercial use and consequent low price at that time.[36] The accumulated zinc at Granby was used to construct a stockade for protecting the women and children against the many local Civil War raids. Following the war this mineral was removed from the stockade walls and marketed at three dollars a ton.[37] Minersville, however, was the earliest heavy producer of zinc in the postwar period.

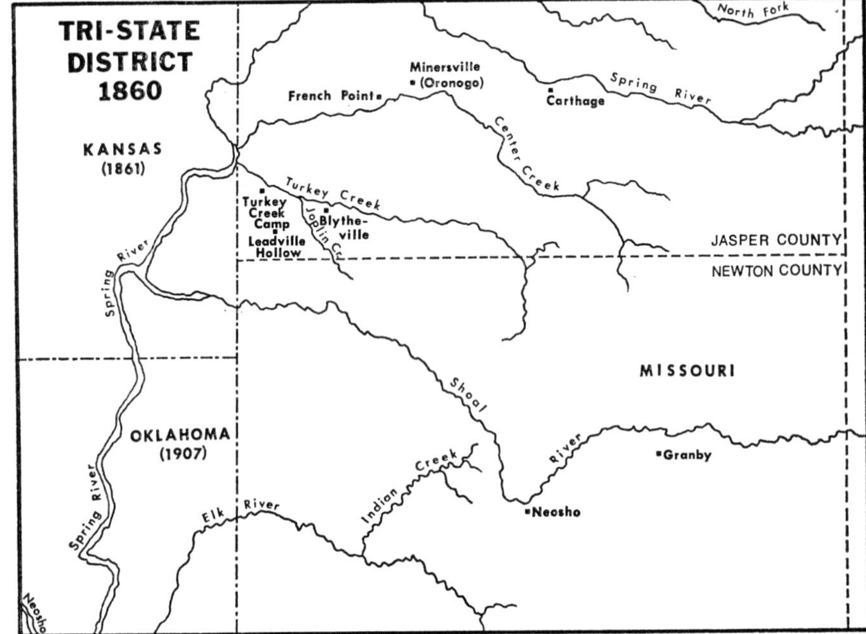

The *Reports of the Geological Survey of Missouri* refer to Minersville as the Center Creek Camp and reveal that by 1855 its mines had produced 419,000 pounds of refined lead.[38] South of these mines several prospects from Newton County came in during 1850. These, too, served as nuclei for mining camps and later for towns. The mining camp of Granby was established late in 1854 with discoveries of lead at the Prairie Diggings, one mile south of present Granby. George C. Swallow made the first state geological survey of the region early in 1854. He reported not a single cabin on the site where Granby now stands.[39] Yet by 1860 there were three hundred shafts around the city,[40] and it claimed a population of seven thousand persons.[41] One of the largest operations in Newton County was conducted by the Blow and Kennett Mining Company, which in 1850 obtained a lease on lands comprising the future site of Granby and belonging to the Atlantic and Pacific Railroad.[42] This company, the only large operator in the district before 1865, carried on mining as well as smelting. Soon Blow and Kennett smelters became so efficient that small operators came to sell their raw ore to this company. Blow and Kennett smelters consisted of six Scotch Hearths, the blast furnished by steam-driven fans. To 1861 this enterprise produced 35,414,014 pounds of refined lead.[43]

Another operator of some importance in Newton County was J. W. Moseley, who discovered lead ore on Shoal Creek between Granby and Neosho in 1850. Within three months his mines produced one hundred thousand pounds of mineral from two shafts at sixty feet with a crew of six men. Moseley's smelting furnace was equipped to handle three thousand pounds of mineral a day.[44] His Shoal Creek mines were such heavy producers that they were still operating in 1894.[45]

While smelting will be treated subsequently, it is appropriate to note here some of its salient features in the pre-1865 period, when the refining process was so closely connected to the mining operation itself. In the first few years of operations each mine had its own smelter adjacent to the shaft. The lead ore in those days was shallow mined. Seldom did the shafts exceed sixty feet, and the mineral was generally 80 per cent or more pure galena; therefore it required little or no concentrating. The ore was hand picked from the debris, hoisted from the mine, washed, and when sufficient quantity had been accumulated, it was placed on a log pile or furnace and the combustible material ignited.[46] The molten lead was poured into molds, cooled, and removed, ready for market. The piece of molded lead, approximately eighty pounds in weight, was called a pig.

The blast furnace was introduced by Moseley near Neosho in 1852.[47] Soon smelting became a specialized operation, and miners brought their raw mineral to nearby smelters where it sold for twenty dollars a thousand pounds.[48] Each camp had its smelter. Kennett and Blow were at Granby, Moseley at Neosho, and the Harklerode Company smelted for the Turkey Creek mines.[49] Specialized smelting involved the use of improved methods. The primitive log furnace was replaced by the Scotch Hearth furnace, which included a blast of air furnished by huge bellows operated by water power or steam. Fuel for the new type of furnace was supplied by charcoal and coal. The lead ash, lost in the log furnace, was rerun through the blast furnace, making extraction more efficient and increasing the profit per ton smelted by 20 per cent.[50]

Aside from the smelting enterprises of Kennett and Blow, Moseley, and Harklerode, the district business organizations remained small. Most of the mines were operated on a partnership basis, with only small amounts of hired labor. This remained the operational pattern until after 1900.

Marketing lead presented a real problem in the pre–Civil War period, since there were no railroad facilities immediately available to the southwestern mining camps. Transportation difficulties were a major factor in limiting the area's full development. The common mode of transporting lead to markets was to haul the pigs in wagons to Spring River and Cowskin, load the metal on flatboats, float to the Grand River in Indian Territory, then to the Arkansas, and on to New Orleans.[51] The *Western Journal* noted in 1851 that three flatboats of lead loaded by G. W. Moseley in Newton County had arrived in New Orleans.[52] Metal shipped by the New Orleans route was generally consigned to New York and Boston markets.[53] Several operators shipped refined lead by horse and ox-drawn wagons to Osceola, on the Osage River. From there the metal was flatboated to St. Louis via the Missouri.[54]

One of the leading proprietors in the Southwest, William S. Moseley, summed up local transport problems in his 1851 article for the *Western Journal and Civilian*:

> The only obstacle to the rapid development of the mineral resources of southwest Missouri is remoteness from navigation, or at least, certain navigation. You are aware that South Grand River flows entirely through the Indian country, emptying into the Arkansas about five miles below Fort Gibson. If that small wedge of land belonging to the Senecas, Seneca-Shawnees, and Quapaw Indians lying west of Newton County was attached to the state it would add greatly to the growth of that part of our state, and rapidly increase the growth and development of one of the finest sections of southwest Missouri. A fine flourishing border town would grow up; steamboat navigation of the Grand River would take the place of road wagons; and thousands of acres, now lying waste, would soon yield up their hidden treasures. From actual inspection and observation, I am satisfied that in the course of time western Missouri will be one of the richest mineral counties in the state. It is the intention of the different mining companies in Newton County to charter a steamer to come up to the mouth of the Cowskin (in the Seneca country) next Spring and carry off their lead.[55]

Mining operations were curtailed considerably during the Civil War, and the camps were desolated by the foraging activities of both Union and Confederate armies. Neosho, near Moseley's camp, was briefly the Confederate capital of Missouri and thereby became the special target of Union operations in the state. The records show that the armies of the North as well as the South were also interested in the region as a source of lead for munitions. The Confederate Army garrisoned at Granby in 1861 and actively operated the mines and smelters throughout the district. Wiley Britton, a contemporary, calculated that the refined metal produced there under Confederate auspices was sufficient to supply the small arms ammunition needs for a large part of the Southern army.[56]

Confederate forces held and operated these southwest Missouri lead mines until October of 1862. On October 11, 1861, the quartermaster at Fort Smith advised Acting Secretary of War J. P. Benjamin, C.S.A., that he could haul 200,000 pounds of lead each month from the mines of southwest Missouri.[57] On October 14, 1861, the quartermaster informed Benjamin that he had shipped 32,000 pounds of lead from the southwestern Missouri mines to Memphis. He added, "I . . . will continue to forward lead and I believe I can furnish all that is wanted for the Confederate Army."[58]

Union forces drove the Southern army out of the lead-producing field in the autumn of 1862 and dominated the area thereafter with production going to the Union Army. Guerrilla forces attacked the Center Creek and Turkey Creek camps in May of 1863, but were driven off by Federal units. Thereafter relative calm settled over this desolated mining field.[59]

With the cessation of hostilities in 1865, the mining region of southwest Missouri was again ready to undertake production for the markets of the world. Aside from the limitations of transportation and lack of adequate population, the region offered every inducement for exploitation and development. S. Waterhouse published a prospectus in 1867 in which he encouraged immigration to the Southwest. There, he said, the prospects for quick wealth were everywhere and lead virtually came to the surface. He advised people of ambition to immigrate to this "sunset side of the Father of Waters."[60]

CHAPTER III

District Development since 1865

Just as the geology of the Tri-State District was a controlling factor in mining and milling methods and land use, so was it of great influence in the founding, growth, and longevity of mining camps and towns. The miners needed centers to purchase mining supplies, food, clothing, and to furnish shelter and entertainment. Limited transportation facilities required the miners to live adjacent to the mine workings. This requirement, added to the widely disseminated ore deposits, gave rise to a multiplicity of mining camps. If the geology was right, the mining camp became a town, but if the deposits exhausted quickly, the camp folded, and merchants moved to new fields in the district where new camps grew up.[1] The more extensive the deposit, the longer-lived the camp; and if a camp promised to furnish profitable mining over an extended period, the camp became a town.[2]

The evolution of Tri-State camps and towns can be divided into four stages of urbanization. First was the mining camp, usually short-lived if the ore deposits beneath its environs were limited. The camp was given a longer lease on life if deposits in the immediate vicinity proved more extensive than at first known. Then a town developed. The mining town, the second stage, was characterized by workers still living in the local area and working the mines of the vicinity. Soon the economy of the town became more diversified. Increased functions included supplying surrounding camps with their mining needs. In the third stage workers lived in the larger towns and commuted to the mines via electric trolley. The smaller towns in this phase, which were bypassed by mining, extended their longevity by becoming farm and trade centers. During the fourth and most recent stage, when 90 per cent of the mining was carried on in the Picher field, some miners lived in Picher or Cardin, but many resided in Miami, Commerce, Galena, and Joplin. The workers commuted by automobile over good roads. More convenient transportation helped more than any single factor to produce this fourth stage in district urban development. A sort of sifting process took place. Each stage of urbanization represented a screen. Many camps failed to pass even the first test; thus today thirty urban centers remain of the original array of camps.

Eighty-one camps existed at one time or another in the Tri-State District. In southwest Missouri these included Lehigh, Wentworth, Pierce City, Carthage, Neosho, Grand Falls, Dayton, Scotland, Leadville Hollow, Chitwood, Carl Junction, Thurman, Webb City, Carterville, Oronogo (Minersville), Smithfield, Waco, Thoms Station, Four Corners, Blende City, Duenweg, Prosperity, Porto Rico, Fidelity, Diamond, Murphysburg, Joplin, Blytheville, Stephens Diggings, Sherwood Diggings, Cox Diggings, Pinkard Mines, Leadville Mines, Carney Diggings, Tanyard Hollow, Taylor Diggings, Belleville, Jackson Diggings, Central City, Saginaw, Swindle Hill, Parr Hill, Lone Elm, Moon Range Diggings, Spring City, Racine, Seneca, Alba, Lawton, Granby, Bell Center, Central City, Asbury, Burch City, Klondike, Georgia City, Stark City, and Cave Springs. Southeast Kansas produced Galena, Badger, Peacock, Empire City, Crestline, Baxter Springs, and Treece.[3] In northeast Oklahoma were Tar River Camp, Picher, Commerce (Hattonville), Douthat, St. Louis, Hockerville, Quapaw, Cardin, Lincolnville, Peoria, and Century.

The origin of the camp names is obscure. Geographical factors were important, as in the case of Grand Falls. District mineralogy produced some names like Blende City and

Galena. Prominent individuals also contributed their names to settlements: Picher, after the mining family responsible for its development; Joplin, after Rev. Harris Joplin; and Webb City, for the owner of the farm where mineral was found. Equally as interesting are the names of mines situated in these camps.

Some of the more famous mines with picturesque names include the Clara Belle, Little Francis, Poor Boy's Mine, Osceola Mine, Big Run Mine, Athletic Mine, Golden Rod and Cornfield mines, Lucky Bill Mine, Lucky Joe Mine, Merchant's Mine, Klondike Mine, Bonnie Belle Mine, Etta May and Betsy Jan mines, Fighting Wolf Mine, Mary Gibson Mine, Sweetman Mine, Silver Plume Mine, Lucky Jew Mine, Sunflower Mine, Bluebird Mine, Rosebud Mine, Blue Goose Mine, Once More Mine, Greenhorn Mine, Navy Bean Mine, King Solomon's Mine, Bonehead Mine, What Cheer Mine, Gilt Edge Mine, Yellow Jacket Mine, Mihomi Mine, Bunker Hill Mine, Netta Mine, Never Sweat Mine, Quick Seven Mine, Aladdin Mine, Osage Mine, Indian Chief Mine, Mahutaska Mine, Redskin Mine, Golden Eagle Mine, and Damfino Damfuno Mine.[4]

Even more obscure than camp names are the sources for the names of the mines. Since the miner was more closely associated with the mine workings than with the camp itself, his names for mines undoubtedly reflected more of his personal life—his hopes and disappointments and such intimacies as sweethearts, wives, and children. Many Tri-State miners had immigrated from other mining fields and brought along the names of previous diggings. After the field of operations moved into Ottawa County, Oklahoma, on Quapaw lands, many mine names showed the influence of Indian lore.

Space forbids a detailed description of all eighty-one mining camps. Suffice it to say that most of the villages and towns of southwest Missouri, southeast Kansas, and northeast Oklahoma owe their existence in the first place to the discovery of lead and zinc deposits in the vicinity of each. Of the eighty-one camps, only thirty remained by 1950. When the minerals in their environs had been exhausted, the extinct camps became ghost communities as rapidly as they had arisen. The remaining camps, serving as nuclei for modern towns, owe their existence to two factors: either a more permanent basis for survival was found, such as agriculture, or the ore deposits were sufficiently extensive to support a community until recent times. The first group includes Carthage, a grain and dairy center, and Neosho, a dairy and fruit-producing community of wide fame. Modern Granby is a farming center.

Webb City and Carterville, the scene of some mining, furnished homes for workers who commuted to the mines of Oklahoma. Also, considerable amounts of mining and milling machinery were manufactured in Webb City. Galena long remained a mining town and was the site of a smelter. Baxter Springs, like Granby, became a farming community. Until recent times Picher, Cardin, and Commerce best exemplified the mining camp.

Of the thirty surviving towns Joplin alone developed into a small metropolitan center, and the rise and development of Joplin illustrates best of all the process of urbanization in the area. In pre–Civil War reports,[5] Leadville Hollow, Center Creek, Turkey Creek, Granby, Neosho, French Point, and Blytheville were mentioned as mining camps. These were largely destroyed by Union-Confederate operations and guerrilla warfare. Quantrill, the most famous of the guerrilla chieftains, cut a swath of ruin and destruction from Leadville Hollow in Missouri to Baxter Springs in Kansas, where he perpetrated his infamous massacre.[6] The resiliency of the region was shown, however, by the rapid resumption of mining operations after 1865.[7]

Numerous prospectuses were published immediately after the war praising Missouri's resources and inviting capitalists and immigrants to take advantage of the state's opportunities.

These included a description of lead ore in Newton and Jasper counties and observed that this mineral was the staple of the area.[8]

The prewar nucleus for Joplin consisted of the Leadville Hollow Camp, the Turkey Creek Camp, and Blytheville. None of these were over two miles apart. Joplin's history in the postwar period begins at Oronogo, where two miners, E. R. Moffett, and J. B. Sergeant, lead miners from Wisconsin, were operating a partnership shaft on lands belonging to the Granby Mining and Smelting Company.[9]

To foster the prospecting and development of its lands, the company had offered a reward of five hundred dollars to the miner-partnership producing the most ore in the four months from March 4 to July 4, 1870. Working in partnership, Moffett and Sergeant were able to collect the reward, and they used the money for a grubstake to prospect in the Joplin Creek field. This area had seen only desultory operations during and after the war, and as yet no miners had found ore in sufficient quantities to start a stampede.[10]

The prewar diggings had caved, scrub oak and sumac had reclaimed the clearings where milling and smelting sites had been located, and in general the valley was a wasteland. Moffett and Sergeant leased a ten-acre tract from John Cox in August, 1870,[11] pitched their tent, the first dwelling in Joplin proper, and commenced prospecting along Joplin Creek. Their prospect shaft, showing rich lead ore, was situated about five hundred feet north of the present Broadway bridge over Joplin Creek.[12]

Before the winter was out, tidings of the rediscovery of mineral in Joplin Creek Valley spread, and twelve prospectors joined Moffett and Sergeant. This number increased to twenty, and by August, 1871, according to Joel T. Livingston, "there were about five hundred people in the camp, most of these men."[13] A companion camp was established in July of 1871 around a body of ore a quarter-mile west of the Moffett-

Sergeant diggings. It was called Murphysburg after its founder, Patrick Murphy.[14] Murphy came from Carthage, purchased a forty-acre tract, laid out mining plots, and organized the Murphysburg Town Company. He shortly built a store and smelter just off the present First and Main streets of Joplin.

In August, 1871, a correspondent of the Carthage *Weekly Banner* was "surprised to find, instead of an expected four or five shafts with a dozen men working, five hundred men and plenty of shafts. Some miners were making $40 to $50 per day." He noted that the general area of the camps was still called Blytheville.[15]

By the end of the year Blytheville and Murphysburg had a combined population of two thousand. This was about equally divided, and from the start an intense rivalry was evident between the camps.[16] Local friction, however, did not materially affect the mining operations and growth of these two camps. By 1874 the area was producing two hundred tons of lead each week.[17] The mineral at this time was found at relatively shallow depths, often in huge pieces. One miner, Alonzo Bradbury, uncovered a four thousand-pound chunk of pure galena at fifty feet. To get this large specimen out without breaking it, Bradbury enlarged the shaft and used special hoisting equipment. Then, with a sled, he hauled the giant piece of galena to Joplin's first hotel, the American House, situated in Murphysburg, and placed it on display.[18]

In the same year that Murphy laid out Murphysburg, John Cox platted a town around Blytheville and the Moffett-Sergeant lease and called it Joplin, after Rev. Harris Joplin.[19] Both camps were regarded as temporary establishments. The dwellings were crude, hastily constructed slab shanties, tents, and covered wagons. According to Dolph Shaner, "Some cooked in the open and slept like soldiers on the march."[20] The area was isolated. Supplies had to be freighted in and lead hauled out in wagons, and the twin camps hardly gave the appearance of a nucleus for a city. But growth was rapid, and by the winter of 1871, Joplin had the following establishments: two hotels, two livery stables, two grocery stores, two dry-goods stores, one hardware store, three general stores, one furniture store, one meat market, one boot and shoe store, a drugstore, a restaurant, a carpenter shop, and five saloons. Murphysburg boasted at the same time one hotel, one livery stable, three grocery stores, three dry-goods stores, one meat market, one general store, one boot and shoe store, two drugstores, three restaurants, one wagon and blacksmith shop, and a carpenter shop, plus three saloons.[21]

Two on-the-scene observers have left the following impressions of early Joplin. F. A. North reported:

> During this time the greatest excitement prevailed, and hundreds of fortune-seekers poured in from all sections. Novices in the use of the pick and shovel, as well as miners of experience set to work, intent on pecuniary gain. The population of the camp had already been swelled to nearly four hundred, most of whom were living in tents while some had hardly so much as a blanket to shield them in hours of slumber from the winds and storms.[22]

Joel T. Livingston recounted:

> The two towns made phenomenal growth and naturally where so many people congregated with no local government everything went. The miners about the camp living as they did in a constant state of excitement, and without the refining influence of the home, for often the family was left behind, plunged into a continuous round of merry-making and the lawless element, unrestrained by officers, had everything their own way, and so bad was the winter of 1871–1872 that it came to be known as the Reign of Terror. The great majority of the prospectors were from Southwest Missouri and Southeast Kansas . . . and while they lived amidst the wild excitement of the day and participated in the revelry they were good citizens. But there were wild

reckless elements there too as in any red hot mining camp. Street fights were common and occasionally the excitement of the day was heightened by a shooting scrap.[23]

After a market for local zinc ore developed during 1872, there was an additional attraction for mining in the Joplin District. Zinc at deeper levels occurred five to six times more plentifully than lead.[24] This further congested the camps and produced added law and order problems. But Joplin was not unlike other frontier communities at a time when lawlessness abounded. Dolph Shaner, who grew up with the two camps, furnished a graphic scene of early Joplin:

> The year '72 started with a big boom in the sale of town lots and construction of homes and business buildings. New mines were discovered and new camps started which made Joplin their trading point. Speculators grabbed land and lots. Mining companies leased more land and employed more miners. Gamblers, women of disrepute, moochers, ruffians, and a generally undesirable class rushed in for easy pickings. As there was no local police and the sheriff's office was in the far off town of Carthage, there was no officer to call to keep the peace. The hoodlums, bad men, and drunks took advantage of such a condition. They showed no respect for decency. Shooting at all hours was a pastime. Windows were targets. The drunks lay around the streets. There was no jail in which to place them and no officer to make the arrest. Saloon fights and street fights were common and there were a few fatal shooting scrapes. A saloon or two closed at noon Sunday but most of them never closed. Women did not feel safe to be on the streets day or night.[25]

The problem of law and order was intensified by the fact that two rival mining towns, Joplin and Murphysburg, existed side by side.[26] Culprits took advantage of this rivalry by committing crimes in one and taking refuge in the other, thereby gaining a sort of asylum. A virtual "reign of terror," according to Livingston,[27] gripped the two camps through the winter of

1871–72. The more responsible citizens of both camps held a mass meeting on February 27, 1872, to discuss the possibility of uniting the camps and gaining incorporation.[28] It is significant that Sergeant presided and Murphy was one of the speakers. On March 14, 1872, the Jasper County Court authorized incorporation of both Murphysburg and Joplin into Union City. A board of trustees was elected to regulate the affairs of the new town. It promptly passed "ordinances looking to the betterment of the town, controlling disturbance, pistol toting, drunkenness . . . and a small jail was built on Broadway between the two towns. The jail was made of two by four oak timber and had two apartments or cells."[29]

Despite these corrective steps and an abatement of the "reign of terror," ill-feeling continued between the two camps and secessionist sentiment was strong. Local pride in each camp was shown among merchants who refused to pay a Union City tax. One businessman, a saloonkeeper, initiated judicial proceedings regarding the legality of Union City incorporation. In December, 1872, the Barton County Court, on a change of venue, upheld the saloonkeeper's contention and ordered Union City dissolved.[30] The union group in the two camps resumed their efforts and through mass meetings and other means were able to accomplish another union on March 23, 1873.[31] Murphy appropriately recommended that the new municipality be called Joplin.

This new town chose E. R. Moffett as its first mayor.[32] Within a year following union, the population of Joplin had increased from four thousand to five thousand.[33] Gradually rivalry waned, law and order were established, and the slab shanties and tents of the Joplin Creek Valley were replaced by permanent homes in platted residential districts. Miners and businessmen no longer hesitated to bring in their families. Schools and churches were erected, and by the end of 1875 there were 1,620 school children with educational facilities. Commercially Joplin had grown too, boasting twelve dry-goods and clothing stores, twenty-seven grocery stores, six boot and shoe stores, five hotels, twelve blacksmith shops, eight livery stables, three lumber yards, sixteen abstract and law offices, sixteen physicians, and two banks.[34]

Yet, despite the growing influence of respectable people in the affairs of Joplin, one so disposed could still find ribald entertainment in any one of many local establishments. Livingston recounts that in 1875 Joplin had

> seventy-five saloons open both day and night and in most of them a full orchestra gave free concerts every night and in most a matinee Wednesday and Sunday afternoons. The following are names of some of the popular bars: Healthwood Bar, Board of Trade, and the Steamboat Salon, the Golden Gate, Miners Drift, Bullock and Bouchers, the Bon Ton, the Palace, and the Brick Hotel Bar. One of the popular places was Blackwells Bar and there something new and exciting was always pulled off. In November 1876, Mr. Blackwell arranged for the entertainment of his patrons a fight between a Cinnamon bear which was brought up from Arkansas and six blooded bull-dogs. One thousand two hundred people witnessed the fight which was won by the bear.[35]

A local miner-poet pointed out that "Suez was still east of us and there were no Ten Commandments, for way down yonder in Southwest Missouri, where women drink and curse like fury; where the barkeepers sell the meanest liquor which makes a white man sick and sicker, where the tinhorns rob you a little quicker, that's where Joplin is."[36] The respectable citizens could entertain themselves through the more sophisticated program offered by the Joplin Opera House.[37]

Granby resented the rise of Joplin, for this Newton County mineral center took great pride in the claim of being the first mining town of the Southwest.[38] Granby's chagrin for the upstart Joplin was heightened by the attraction of large numbers of her miner residents to Joplin, and she showed her contempt by referring to Joplin as a camp and "not a town."[39] The social ecology of Joplin was of an encroaching and absorptive sort, which could well invite the indignation of sister camps, and its population grew from immigrant miners and the annexation of peripheral camps.[40] A study of surrounding camps shows the assimilative nature of Joplin. Chitwood, Blende City, Thousand Acre Tract Camp, Moon Range, Swindle Hill, Turkey Creek, and Lone Elm all shortly became a part of Joplin's city limits. The annexation of Lone Elm alone brought 2,500 people into Joplin.[41]

New discoveries at Webb City, Carterville, and Galena around 1876, and at Picher, Oklahoma, in 1914 started another miner stampede and partially depopulated Joplin for a brief period. Many miners from Joplin moved into these new fields and helped develop them, but their places at Joplin were filled shortly with newcomers. Rather than decreasing the importance of Joplin, these new camps actually improved her position.[42] Interurban electric trolleys enabled many of the miners to live in Joplin, commute to the mines, and provide their families with the advantages of a larger community. Four years after the opening of rich deposits at Galena, Webb City, and Carterville, Joplin's population had increased to eight thousand,[43] and in forty years this had grown to nearly five times the 1880 figure. As North put it, "Galena, Empire City, Thurman, Webb City, and Carterville, Oronogo, Smithfield, Carl Junction, and Blende City, are little more than suburban towns, each paying royalty to the great city, Joplin, around which they cluster and upon which they are in a large part dependent."[44] Just as Joplin absorbed its peripheral camps through annexation, the "town that Jack built"[45] also assumed those basic distributive functions that supplied the more distant camps. More and more Joplin was diversifying its economy, and a disease "called 'Joplin Panic or Colic,' characterized by being head over ears in various enterprises, seemed to have seized the inhabitants."[46] Several promoters referred to Joplin as the Klondike of Missouri, but its wealth was rapidly being derived from sources other than mining. By 1895 several wholesale businesses were exporting food, clothing, and mining supplies to the camps in Kansas and Oklahoma. This included two liquor wholesale houses.[47]

The Joplin Colic was best exhibited on Saturday night. The town's five banks were open until 8:00 P.M. to accommodate customers. Mine operators, land owners, workers, and ore buyers assembled in the banks for the weekly payoff. Grocery, dry-goods, general, and hardware stores operated until 9:00 P.M. to meet the needs of miners and their families who crowded the streets. Saloons remained open until the thirst of their patrons for drink and merriment had been slaked, and "the theater was crowded, for Joplin patronizes theatrical entertainments lavishly in its pretty Club Theater."[48]

As noted earlier, the genesis of Joplin, especially its first two periods of mine urbanization, was repeated eighty times over in the other camps. While space limitations forbid a detailed description of each of these, some attention must be given to the Oklahoma camps, since this field became the district's heaviest lead and zinc producer. The discovery of ore there came rather late. The official reports make only slight mention of lead mining in present Oklahoma before 1891.[49]

Several unofficial references can be found, however, which indicate a knowledge that minerals existed in present Ottawa County and were exploited in a limited fashion. G. W. Moseley wrote in 1855 that "our mineral region embraces the counties of Jasper . . . and Newton in Missouri, and the country claimed by the Quapaw, Seneca, and Shawnee Indians in the Northeast Portion of the Cherokee Country."[50] H. D. Andrews, a former Quapaw Indian agent at Miami, noted that shallow lead was mined for shot by Quapaws as early as 1870 near present-day Peoria,[51] and North described several exploratory expeditions out of Joplin into northeastern Indian Territory.

One of these was a hunting party led by a mining operator from Carterville in 1878.[52] Broadhead's *Report for 1873* discloses that a miner named Sibley sank three shafts, each producing lead, one and one-half miles south of Seneca, Missouri, near the line of Indian Territory.[53] Commercial lead and zinc mining was first carried on in the Oklahoma portion of the Tri-State District in 1891. The Peoria Mining Land Company, a New Jersey corporation operating out of Joplin, was chiefly responsible for developing the new field. Lead and zinc concentrates from Peoria were hauled by wagon to Joplin and Baxter Springs, since Oklahoma's first mining camp lacked a railroad.[54]

The productive period of Peoria ranged from 1891 to 1903, and the value of minerals extracted from her mines was $91,068.[55] Lincolnville was the next camp to develop in Indian Territory. A farmer, excavating a well, unearthed a heavy vein of lead ore near there in August, 1897.[56] By 1903 a camp of 1,200 miners had been established, and prospecting showed that the ore run extended westward for about two miles in an intermittent fashion.[57] On the western edge of this deposit, another mining camp, Quapaw, grew up in 1904. The Quapaw area remained the principal producer for the Oklahoma district until 1914, the value of its output exceeding $1,042,000.[58]

Meanwhile, a camp opened over a deposit three miles north of Miami in 1907. Sometimes referred to as the Miami camp, actually the camp's official designation was Commerce, although for the first two years it was called Hattonville. Here, as in the case of Peoria, Lincolnville, and Quapaw, miners from Joplin and Galena moved in and applied mining and milling techniques characteristic of the more northerly portion of the Tri-State District. In many instances Joplin operators actually dismantled hoists and mills from Jasper County mines and reassembled them in Ottawa County.[59] From 1908 to 1914 the Commerce field produced $4,798,286 worth of lead and zinc concentrates.[60]

The last big strike made in Ottawa County, or in the entire district for that matter, was at Picher. The Picher Lead Company of Joplin had been prospect drilling on its lease near Commerce. The results were disappointing, and the drill rigs were ordered back to Joplin. One rig, en route, mired down near Tar Creek in August, 1914, and while waiting for relief, the driller sank a wildcat hole. The cuttings showed rich ore signs. Other company rigs moved in and a drilled tract was blocked out, each hole indicating heavy deposits. The Picher Lead Company obtained a lease on 2,700 acres in the prospect area, and a new camp, Picher, grew up overnight.[61] The Tar River area, before the wildcat prospect, was occupied only by isolated farm houses; but within a year after the discovery, Picher and Cardin each had a population of from 750 to 1,000.[62]

Thus the mining center of gravity shifted to Picher. Ninety per cent of the district's production after 1915 came from this field. The Tri-State Ore Producers Association, an operator grouping, had its headquarters at Picher. A local of the United Steel Workers of America, CIO, bargaining agent for the district's unionized miners, also was situated there. But the financial, wholesale, and distributive center and the equipment manufacturing and supply source continued to be Joplin, the nerve center of the Tri-State District.

CHAPTER IV

Tri-State Prospecting

In the early days of the Tri-State field the mining process consisted of prospecting, mining proper—the excavation of the ore and debris from underground workings, milling—the separation of lead and zinc ore from the dirt and rock debris, and smelting—refining the milled ore into metal. From 1848 until 1865, most, or in some cases all, of these operations were carried out by the miner at the site of the mine. The word *mining* meant that one carried on all of these phases. Gradually these mining steps were separated, until each operation became highly specialized, technical, and mechanized. Likewise, these operations were isolated in that specialists, each trained for a particular phase, carried out their functions in a separate fashion, often apart from the mine itself, as in the case of central milling. Mining came to refer solely to the extracting of ore from underground workings.

Prospecting was the process that determined the location and probable extent of ore deposits. The methods used in this vital step of mining ranged from legerdemain, consulting fortune tellers, and use of the willow switch, to geophysical exploration.[1] The first local prospecting undertaken in 1848 was accomplished by sinking a shaft, too often put down blindly.[2] Some prospectors with rudimentary geological knowledge paid close attention to the connection of topography and ore deposit location. H. F. Bain and C. R. Van Hise admit that this is often a reliable guide, but that "the unschooled may push it too far."[3] Sinks or synclines in valley floors were favorite sites for shaft prospecting. The dimensions of these test pits varied from four by four feet to four by five feet. Seldom did the depth of the prospect shaft exceed fifty feet.[4]

Pumping facilities were largely absent, and since ground water was encountered about fifty feet down, the shaft was abandoned if no ore was found to this point. The older portion of the district is dotted with these abandoned shallow shafts and the accompanying pile of dirt and rock, "silent monuments of discouragements and disappointed hopes of the miners, each of whom in opening them pictured to himself finds of rich ore and rapidly accumulating wealth."[5] A hike through the hills in any direction from Joplin will disclose dozens of these incipient mines to the square mile.

As the test shaft went down, the miner would lower himself in a bucket suspended on a hand windlass operated by his mining partner. With pick, shovel, pry bar, and light powder, when heavy rock was encountered, the underground worker would gouge and probe for ore, shoveling the dirt and rock into the bucket. This was raised on the windlass and dumped on the debris pile. The quantity of ore found to the water level determined whether the prospect was abandoned or developed into a mine.[6]

When the churn drill came in for wide use around 1900, deeper explorations on the brink of these shallow prospect holes often revealed extensive ore deposits immediately below the abandoned diggings. By this time the shallow deposits were largely exhausted, and mining had to extend below the water level.[7] Thus shaft prospecting became too expensive, and of necessity the churn drill was introduced.[8] Even before the entry of the standard mobile churn drill, several attempts were made to get away from the slow, and often unproductive, prospect shaft system. These centered around the use of stationary drills with a churn or lift-and-drop action.[9]

The carpenter's rig was the most common between 1890 and 1900. It consisted of a platform and drill tower built over the anticipated deposit. This supported the pulley over which the drill rope passed. One end of this rope was attached to the drill tool, the other end to a drum which raised and lowered the rope as the tool head cut deeper. As the drum was turned, either by horse, ox, or steam power, the driller would raise and lower the cable by operating the drum brake.[10] In this fashion the heavy drill point was driven slowly into the earth, its cuttings examined from time to time for signs of ore.

While this crude drilling device generally indicated the geological character of the ground below the levels to which prospect shafts could be sunk, its stationary character limited the drill's effectiveness. Once mineral trace had been found, several prospect holes had to be cut in order to determine the extent and thickness of a deposit. This meant dismantling the carpenter's rig, moving the pieces to a new location, and reassembling the unit before drilling could resume.[11] To overcome this handicap, miners began around 1900 importing the mobile churn drill from the nearby oil and coal fields.[12]

This early portable drill was moved by horses or oxen and differed from the carpenter's rig only in that it could be moved intact. Later models were mounted on units powered by steam or gasoline engines. Capacity depth of the self-contained, portable drill, known locally as drill rig, was from 150 to 250 feet. In 1900 a Star or Keystone rig with one set of tools cost around $1,100. If the rig were a nontraction type, it required four to six horses to move it to a new location. The drill unit was powered by steam engine, mounted on the rig frame, and fired with bituminous coal at a cost of $1.50 to $2.50 for each ton delivered. Water for the boiler and drill-hole washing was brought to the rig in barrels and tank wagons. About two barrels of water were required each hour at a cost of from five to eight cents a barrel.[13]

A large drum with clutch and brake attachments, turned by steam power, raised and lowered the drill rope that passed up to the boom tower and thence vertically to the drill position. A line of tools, the excavating device, was attached to the drill rope. The tool line consisted of a heavy bit, auger stem, sinker bar, and rope socket and had a combined weight of 1,800 pounds.[14] This formed a head sufficient to drive through soil and rock when dropped in short, churning thrusts. As the hole deepened, lengths of four-inch round iron rods were added to the tool line. Most of these early portable rigs were adapted to horsepower if the steam engine failed.[15] Two men were required to operate the drill rig, one to run the drum lift, the other to guide the tool line. The latter's chief duty was to turn the line one-quarter each stroke to insure a round hole. The diameter of the drill hole produced by the churn unit was 5⅞ inches.[16]

Cavities and fissures often led the tool line off-center and, unless corrected, caused trouble during casing. To remedy this, it was common practice to remove the line of tools, lower a charge of dynamite into the cavity, and blast the opening; then drilling was resumed by releasing the drum. The tool line was lowered to within five or six inches of the bottom of the hole, quickly raised, and allowed to drop to the bottom, the force and weight of the thrust cutting the rock and soil. Thirty-five to forty strokes a minute were possible in limestone, while the average in flint or chert was forty to forty-five strokes. The drilling average each day was twenty feet in limestone and six to seven feet in solid flint. An over-all average of ten or fifteen feet a day was considered good.[17] When the drill had progressed to solid rock, the hole was cased with cast-iron pipe up to the ground surface. This insured a vertical thrust and prevented caving around the tool line as it deepened.[18]

In tracing out a deposit, holes were sometimes drilled only fifty feet apart, while in other places, three to four holes an acre were considered sufficient. The Miami camp had drill holes fifteen to twenty feet apart.[19] In 1908 drilling costs averaged ninety cents a foot, and the depth averaged 175 feet, thereby running the cost of exploring an acre to approximately five hundred dollars.[20]

Sheet ground required much less drilling than run-type formations, since its blanket deposits were more uniform, and, as a general rule, two holes an acre sufficed.[21] Shallow drilling, common before 1915, was only slightly less uncertain than pit prospecting. A drilling contractor of long experience commented that with this type of shallow prospecting, "about one hole in ten strikes shines of jack, one in twenty-five fair ore, one in fifty strikes it rich. Some localities seem to be underlaid by an almost continuous body of ore and nearly every drill hole strikes ore. Other tracts are barren for wide areas, or contain ore scattered in infrequent deposits."[22]

A truer picture of the underground horizon could be achieved only by deepening the drill holes beyond the 300-foot level. Where the deposits were parallel at different levels, as in the Miami camp, this was especially true. In 1912, drill holes extending below a shallow strike at 200 feet showed ore at 300 to 350 feet to be much richer than the upper deposits.[23] Once the drill struck ore, samples of cuttings were taken every two feet. The cuttings, collected from the tool line in a powder box, were emptied into a half-barrel or bucket containing water. After the samples had settled, they were assayed to determine the ore content.[24] While some drillers appraised the value of the ore by eye and experience, others used the hand jig in this stage of prospecting. If the drill cuttings indicated a small amount of ore, the driller simply recorded "a few shines of jack" or lead, and if the ground appeared productive he wrote "good shines of jack" or lead.[25]

The churn-type drill remained the chief prospecting device used in the Tri-State District. While this drill rig became more mechanized and self-contained, drilled deeper, and was more efficient, basically it remained the same. Records of drill holes, especially, improved. In the early days drilling contractors frequently failed to log barren holes at all. Later all holes, barren as well as productive, were carefully described. In 1910 the St. Paul Company began to use a more adequate record system, and gradually its practice was universally adopted. This company's method was to require a drill report, made in triplicate, on each hole. Copies were filed in various company offices, and in case of fire or other mishap, at least one drill record was available. This report contained the drilling cost, geological formation, position, texture, water, and ore assay.[26]

Since the mine and mill discharge sometimes covered a quarter-mile square, drill holes, and thereby valuable mineral locations, often were lost by covering them with mill tailings and debris. To avoid this loss, it became common to drive posts, four and one-half inches by 10 feet, into the drill hole and tag them with copper identification plates.[27] Even with the most careful assay of drill cuttings, the churn drill could mislead the prospector. The internal jar of the tool line and wash of water often caused minerals to fall from the upper levels and concentrate in the hole bottom, thereby producing a false record.[28]

Prospectors found that by lowering a short length of sheet-iron pipe loaded with dynamite into the ore bed, exploding it, then drawing out and assaying the material, they could achieve a more reliable reading on the deposit.[29] As a general rule mine operators could count on the ore milling out 10 per cent higher than drill cuttings assayed.[30]

Sometimes even deeper drilling and improved records failed to furnish adequate information on lower run-type deposits. This handicap was partially overcome, especially in the Miami field, by dismantling churn drills, reassembling them underground at the 200- to 250-foot level, and resuming the prospect to deeper levels.[31] Prospecting underground was done by means other than the churn drill. Prospect or pull drifts frequently were driven laterally from the shaft or mine drift to prove the presence of a disconnected ore run which showed on the field's drill log. The dimensions of the pull drift were seven feet by seven feet in cross section. If the run was located by this method, the pull drift became a connecting link to the main shaft by laying track for ore cars. Commonly a contractor furnished the labor, equipment, and powder much as in the case of churn-drill prospecting. The average cost per foot of excavating a pull drift was $7.50.[32]

In recent times a frequent preliminary to driving the pull drift was horizontal drilling with an air-hammer drill mounted on a heavy tripod. This tool probed laterally beyond fifty feet, and its cuttings were examined, as in the case of the vertically driven churn drill, to determine if the pull drift was justified.[33] Because the ore runs were highly irregular, mine prospecting and workings often took on a maze appearance in cross section, quite unlike the orderly prospecting, drifting, and stoping galleries of Western mines. But the inherent nature of the Tri-State deposits required that the miner-prospector adhere to a single rule—namely, "follow the ore."[34]

New prospecting developments, in addition to improved records, deeper underground drilling, and pull drifting, included attempted use of the diamond drill, as well as an increase in the size of the drill hole. Deeper surface drilling, entire field prospecting-development and leasing, and geophysical investigation combined to remove additional risks from mining.

The diamond drill was used successfully in southeast Missouri, where the ground was uniform in texture and structure. Many prospectors believed that it would work equally well in the Tri-State District. But after a few attempts the diamond drill was discarded as a prospecting device in the southwest. This region's seamy ground, containing crevices and broken cherty material, made diamond drilling difficult, and the uneven fissured strata tore out the diamond points.[35]

Improved drilling tools made larger holes. In the 1950's the standard drill hole was 6¼ inches in diameter.[36] Discovery of ore at 300 feet at the Badger and Klondike camps near Galena, Kansas, in 1913, set off a campaign of redrilling the entire field at depths below the 250-foot level. Later, surface drilling was carried on below 400 feet.[37] Mining and land companies exemplified by the Granby Mining and Smelting Company established the practice of buying or leasing large tracts of mineral land. These enterprises blocked out ore deposits through drilling and then leased or subleased mining plots on a royalty basis. If the royalty on the land was 10 per cent before development by drilling and proving, it subsequently increased to 15 or 20 per cent. Early in the development of the Miami camp, the Commerce Mining and Royalty Company prospected deposits and proved the field before leasing.[38]

Several geophysical explorations were carried out in an attempt to prospect the district further. The North American Exploration Company of Houston, Texas, made a brief gravity survey in 1926. Two years later the Radiore Company of Los Angeles, California, undertook an inductive radio-frequency survey to determine the location and trend of underground horizons. In 1933 the Missouri Bureau of Mines and Geology made a magnetic survey of the Tri-State District. Then in 1941 the University of Kansas Engineering Experiment Station and the State Geological Survey conducted an extensive geophysical investigation of the Tri-State District in the hope that a cheaper method could be found for prospecting ore deposits. The gist of these surveys was that little correlation appeared between magnetic deviation and mineral deposits.[39]

Sixty years ago a miner could shaft prospect for lead and zinc at the cost of his and his partner's labor. This method was relatively cheap, but it was highly inaccurate. In recent times highly trained consultants and expensive machinery predominated in the prospect field. This was a part of the general mining trend toward concentration of operations and consolidation of holdings. Before expensive mine and milling machinery could profitably be erected, there had to be assurance that sufficient ore existed to warrant erection. Oliver W. Keener's study of the Hartley-Grantham Mine near Baxter Springs, Kansas, developed in 1931, showed that

> prospecting was done entirely by churn drilling, and forty-three holes were drilled ahead of a proved ore body on the adjoining lease to the north and were concentrated along the estimated trend. Later some cross sections were drilled at right angles to the strike and some general drilling was done, but outside the channel of the main ore body these holes proved to be barren or of no commercial value. The extent and grade of the ore body was well defined by this method of prospecting.[40]

Charles F. Jackson and associates studied the trend to concentration of operation, especially in prospecting, and found that

> it is impossible to forecast the number of holes required to determine whether there is or is not enough ore to warrant opening a mine. Whereas some properties have been under-drilled before a mine is opened, complete drilling of a tract which sometimes requires as many as 250 holes, involves heavy capital expenditure and high interest charges thereon over the life of the mine. Roughly, the rock tons of ore proved before the investment required for opening a mine is made, should be 300,000 to 350,000 tons or expressed in terms of concentrates 15,000 to 25,000 concentrate tons. From 60 to 100 holes may be required to prove the necessary tonnage, which would mean 15,000 to 25,000 feet of drilling. Thus to explore a forty acre tract one should expect to spend $35,000 to $40,000 for drilling, sampling, assaying, supervision, and consultation. . . . In the past the number of failures was multiplied because preliminary exploration was insufficient and it was undertaken and mining operations begun by inexperienced people. . . . Much drilling only proved that large areas were barren of ore, and $50,000 to $75,000 apiece was spent by various companies in prospecting in the heart of the Picher field without finding com-

> mercial ore deposits. . . . Actual records show that $1,500,000 was spent during a twelve year period for drilling 6,690 prospect holes in Cherokee County, Kansas, on properties that were abandoned without any ore being found. An additional $3,000,000 was spent for 9,700 holes on properties which later were brought into production but which have not yet paid back the investment and may never do so. However, many mines have been opened in consequence of results of churn drilling and have returned handsome profits.[41]

In spite of the trend to concentrate operations, small operators continued to use early-day methods. Roy Moore, a veteran small, independent operator, reported that he struck paying ore at eighteen feet near Spurgeon, Missouri, in 1941, using a prospect shaft. Earlier mining, however, had proved the area mineralized.[42]

Clarence Brown, another miner, owned a Ford truck with a churn-drill rig mounted on the truck chassis. At the time of the interview in 1953, he had contracted to drill a water well near Joplin in a mineralized area of heavy production during World War I. This section had been abandoned in the 1920's in the belief that it had been mined out. Literally from the top of the ground to fifty feet, the point of drilling at the time of interview, the driller announced he had found "good shine" and was contemplating converting the water well into a lead mine.[43]

Lone Elm Mining Camp, Missouri. (*Courtesy Joplin Mineral Museum*).

Blende City Mining Camp, Missouri. (*Courtesy Dorothea Hoover*).

Turkey Creek Mining Camp, Missouri. (*Courtesy Dorothea Hoover*).

Carterville Mining Camp, Missouri. (*Courtesy Joplin Mineral Museum*).

King William Mine, Duenweg Mining Camp, Missouri. (*Courtesy Dorothea Hoover*).

A four-thousand-pound piece of pure galena from the Bradbury Mine, on exhibit at the Murphysburg-Joplin Mining Camp, Missouri. (*Courtesy Dorothea Hoover*).

Opening an ore vein at the three-hundred-foot level underground, Netta Mine, Picher Mining Camp, Oklahoma. (*Courtesy Bizzell Library, University of Oklahoma*).

Primitive horse hoister, Tanyard Hollow Camp, Misso

Courtesy Bizzell Library, University of Oklahoma).

Commerce Mining Camp, Oklahoma

ourtesy Bizzell Library, University of Oklahoma).

Churn-drill prospecting, Oklahoma. (*Courtesy Bizzell Library, University of Oklahoma*).

Churn-drill prospecting, Missouri. (*Courtesy Joplin Mineral Museum*).

Webb City Mining Camp, Misso

Joplin Mining Camp, Missou

urtesy Joplin Mineral Museum).

urtesy Joplin Mineral Museum).

Eagle Picher Mining and Smelting Company, Empire City, Galena Mining Camp, Kansas. (*Courtesy Kansas State Historical Society*).

CHAPTER V

Tri-State Mining Methods: The Early Period

After locating a deposit, whether by shaft or churn-drill prospecting, the miner was ready to begin mining proper. If he had used the pit-prospect method, the miner started drifting horizontally, that is, excavating the ore deposit at a right angle to the shaft. If the deposit had been found with a churn drill, his first step was to dig a shaft vertically into the ore bed, then follow the ore by drifting. The methods used in exploiting these deposits were remarkably crude from the beginning and improved only gradually. The Tri-State District was from a generation to a half-century behind eastern Missouri in lead and zinc mining methods. For example, steam for hoisting, pumping, and other uses was adopted in the eastern Missouri mines as early as 1844.[1] In the Tri-State District many mines lacked steam power in 1900, and in the 1950's an occasional hand windlass over a mine could still be found.

This slowness in adopting new methods is attributable to a number of factors. The deposits of the Tri-State District were scattered, and most mines short-lived. Few mines supplied sufficient ore to justify operations for ten years, and rarely did any exist over a twenty-year period. The inherent nature of the deposits and the risk involved discouraged the installation of modern equipment and methods on any grand scale.[2]

Around 1875 several eastern Missouri operators came into the district. Undoubtedly influenced by the records of such famous diggings as their local Mine La Motte, which had been in operation since 1720,[3] these newcomers set up the same efficient, expensive operations characteristic of eastern Missouri, where deep, uniform deposits justified such an undertaking. Very shortly, and at considerable loss to investors, it was demonstrated that the geology of the Tri-State District required a different approach.[4]

Another factor which contributed to the slowness in adopting new methods is best explained in terms of precedent. From 1848 on, the Tri-State District was popularly known as a "poor man's camp."[5] The shallowness of the deposits and the lack of mechanized competition enabled men with little or no capital to become mine operators. Landowners and mining land and royalty companies, aware of the risk involved, divided tracts of mineralized land into mining lots, generally two hundred feet square.[6] These plots were leased to miners in return for the payment of a fixed royalty once a strike was made.[7] Most of the risk was taken by the miner, and the system proved so profitable for landowners and land and royalty companies that there was a persistent reluctance to change.[8] Such a system kept operations on a smaller scale and thereby discouraged the adoption of more efficient methods that necessarily would have come if the scope of operations were larger.[9]

By 1900 mining land and royalty companies were entering the field as operators. This development produced concentration of holdings and more efficient operations. Even this had limits, however, for landowners, who leased their lands to mining land and royalty companies, generally stipulated that all ore raised from a particular shaft be milled at that point. This requirement reduced the possibility of mining companies' pirating the landowner's royalty share. This, too, checked consolidation and centralized milling and kept operations on a relatively small scale, since a "mill-per-mine" requirement meant the tonnage of ore mined for subsequent milling must be limited to the capacity of the mill.[10]

Partnerships predominated in the Tri-State District until 1890. As will be demonstrated, mining there required at least two workers; teamwork was indispensable to successful mining. Few instances can be found of hired labor before 1885.[11] The partners shared the work, misfortunes, and profits, if any.[12] There is recorded an instance in 1876 of a nine-man crew working a mine near Joplin, all of whom were partners with no hired help. Their mine was situated in the fabulous Kansas City Bottoms, where veins of solid lead, sixteen to twenty-four inches thick, ran horizontally from within a few feet of the surface to 110 feet. This nine-man partnership divided labor so that one man was on the windlass hoisting, six men were in the ground, blasting and shoveling, and two workers sluiced the ore on top. Their labor produced fifteen thousand pounds of lead a week, or about forty-two dollars a man.[13]

Two-men associations, however, were far more common. Their mining tools included a double-pointed pick, a square-point scoop shovel for loading the ore can, and a light, round-pointed shovel for digging. A windlass equipped with a rope and bucket served as the pioneer hoisting apparatus. The windlass consisted of a five-foot-long wooden cylinder, eight inches in diameter, fitted with a hand crank. The cylinder rested and turned on two two-by-eight-inch notched uprights, four feet high. Locally the windlass unit was built on a platform directly over the mine. A topside worker turned the crank on a signal from the worker below and raised the 150-pound laden ore bucket, and manhandled it to the ore screen.[14]

If the mine developed as hoped, the operators replaced the crude windlass with a whip hoister. From the adjacent timber-clad hills, the miners cut three oak poles, about six inches in diameter and twenty feet long. These were placed securely in the form of a tripod over the mine opening. A swivel pulley was attached to the top of the tripod. One end of the shaft rope, equipped with a hoist hook, ran through the pulley into the shaft. The other end of the rope was attached to the harness of a team of oxen or horses.[15] This device was called a whip hoister since the hoisterman drove the animals away from the shaft with a whip, thereby drawing the laden ore bucket from the mine shaft. Adjacent to the hoist were two hoppers or chutes. One ran from the shaft to a sluice box where the ore was washed free of rock and dirt. The other led to the debris pile where nonmineralized dirt and rock were dumped. The hoisterman detached the ore bucket, dumped it in the appropriate chute, and reattached the container to the hoist hook. Then, to lower the empty bucket into the shaft, the team was driven back towards the mine.[16]

This operation was much faster than the use of the windlass, and heavier ore buckets of five-hundred pound capacity could be hoisted. A modification of the whip hoister was the horse whim. This type of hoist was constructed by attaching a horizontal sweep-pole, about twenty feet long, to a vertical pivot and set of gears. These turned a drum containing the hoisting rope. One animal could power the sweep, the hoisterman still driving the animal in a circle with a whip.[17]

In 1876 several mines adopted a hoist powered by steam. A single steam engine could power a hoist, pump, crusher, jig, and ventilator fan. The steam hoist was the most efficient type of lift possible for the times.[18] Fuel for the steam engine was furnished by coal from the nearby Kansas fields and wood from the abundant timber adjacent to the mines.[19] The steam hoist could raise as much ore as the cans could hold, about six hundred pounds, in a faster and safer fashion than the whip or windlass hoist. The steam hoist was housed in a derrick building and the operation was protected, whereas the other types of lifts and the workers were completely exposed to the elements. The steam lift was powerful enough not only to raise but also to dump mechanically the six-hundred-pound ore cans down the ore hopper as rapidly as they were filled below. The derrick housing was constructed on heavy stilt framework about fifteen feet above the shaft. The purpose of elevating the hoist shed above the ground level was to enable the hoisterman to dispose of rock and dirt debris efficiently. If the hoist were on the ground level, the debris would soon litter and crowd the mining operation and constitute a safety hazard. An elevated runway carried the nonmineralized material to a point beyond the immediate vicinity of the shaft and dumped it in an orderly, nonobstructive fashion.[20]

In opening a mine, as long as dirt and small rock were encountered, one man worked underground, filling the ore can, while his partner hoisted and dumped on top.[21] It was customary to change off from time to time, the underground worker taking over the hoist, while the erstwhile hoister became a shoveler. The underground worker loosened the overburden with the pick and shoveled the material into the can, gave the hoist signal, and waited for the container to return. The dirt and rock were searched for ore sign, and small quantities were picked out by hand. As soon as heavy or solid rock was encountered, dynamite was used to blast and loosen the ore and debris. If large, soil-embedded boulders were excavated and were too large for hoisting, but yet an obstacle to continuing shafting and drifting, it was common to place powder around the boulder and "pop" it.[22]

When solid rock was encountered, the only alternative was to blast the obstruction loose, and an essential preliminary was to drill holes for powder receptacles. Since this was a two-man job, the topside worker lowered himself in the ore can to assist in driving the powder holes.[23] A hand drill, heavy steel about one inch in diameter, locally called a spud, and a three-pound hammer were the essentials for this operation.[24] One miner holding and turning the steel drill while his partner struck it with the hammer produced powder holes in solid rock twenty to thirty inches deep and spaced twenty-four to thirty-six inches apart, depending on the nature of the rock. Two types of powder were used. The most common and inexpensive in the first few years was regular black blasting powder.[25] Giant powder, while more expensive, was also more effective and gradually came to displace the weaker black powder.

Giant powder, which looks like brown wax, consists of nitroglycerin mixed with either sawdust or sand to dilute its power. This charge was enclosed in a cartridge, sealed with a detonating cap, and attached to a rubber-coated firing fuse that assured sparking even in wet or damp ground. Either type of explosive was fitted with a fuse long enough to enable the miners to station themselves in a safe position before setting off the charge. A common complaint against the giant powder, not found in connection with the milder black material, was that it produced nausea and severe headaches among the miners who used it.[26]

The fact that miners failed to wet down the rock dust after blasting caused a high incidence of silicosis or miner's consumption, locally called miner's con, a lung disease.[27] Only in later times were preventive steps taken to overcome the health hazards of blasting.

The market price of ores influenced the amount of blasting undertaken, the depth of shafts, and the length of drifts. Before 1872 lead was the only mineral sought. Zinc prices were as low as three dollars a ton in 1872.[28] This rate could scarcely pay the transportation costs, and, as a result, zinc ore was thrown out on the debris pile.[29]

Deep shafting was impractical because lead was the only marketable ore, and at upper levels, around fifty feet, it predominated. Below this point, which was also the water line, the proportion of zinc to lead increased until, at three hundred feet, it ran about five to one in favor of zinc. Shortly, however, the price of zinc started upward. Zinc ore, delivered at the rail point in Baxter Springs, Kansas, in 1873 brought eight dollars a ton. By 1879 this had increased to twelve dollars; twenty-one dollars in 1886; and in 1888 zinc ore brought twenty-seven dollars a ton.[30]

This consistent rise in the price of zinc ore after 1873 made it worthwhile to mine both lead and zinc, for with increased receipts from the zinc added to what the miner had always received for lead, he could afford to buy costly pumping, hoisting, and milling equipment.

In his quest for lead and zinc ore, the miner found shaft sinking the slowest of the mining steps. The nature of the ground determined the rate of excavation, the type of tools, and the amount of powder required. In dirt and clay, six feet a day were possible. Cherty limestone or solid chert might slow miners to a rate of one foot every three days. On the average, two men could cut through ten feet a week.[31]

From the collar of the mine to the solid rock, the shaft was cribbed; that is, it was lined horizontally with four-by-six-inch timbers in a cross-tie fashion. Native oak lumber was used for cribbing.[32] Several of the mining land and royalty companies operated saw mills to supply lessees lumber for cribbing, underground supports, and general mine building needs. Many early miners cut their own crib pieces from the district's heavily forested hillsides, hewing and notching them by hand. The purpose of cribbing was to create a smooth surface for the upward-downward passage of the ore bucket and to prevent caving as the mine deepened into the hard rock. A vertical shaft was essential to opening the mine, for the terrain of the district was relatively flat and the ore deposits lay at varying depths below, parallel to the surface. The shaft gave access to the deposit and afforded a means of transporting men and supplies into the ground and lifting ore and debris out.[33]

The early shaft dimensions were four by four feet, allowing sufficient space to lower men and supplies and lift the ore cans. The introduction of improved pumping equipment required a change in shaft dimensions. In order to clear the pump line, a shaft four by five feet became general.[34] Later steam and electrical lines, used to operate underground power machinery, required even larger dimensions to afford clearance. Larger shafts were also necessary to clear dismantled churn drills and permit the entry of mules used for underground haulage. The shaft was sunk to the ore deposit which showed in the debris and on the sides of the shaft as the mine deepened. Once ore was struck, the shaft was deepened about five feet to produce a sump. Into this, water collected from the upper levels, and into it was placed the pump line.[35]

Water was a perennial problem in the mines. Several good mines had to be abandoned from time to time because of underground circulation through the crevices and the consequent filling of shafts and drifts in quantities beyond pump capacity. The early miner had to lift the overnight accumulation from the shaft to the sump level each morning before he could go underground. Buckets lifted on the windlass or horse hoister were the only pumps when the field first opened.[36]

By 1873 the most common mine pump was the Cornish Horse Pump, locally called the walking beam pump. A vertical wooden shaft fitted with a cam movement and center-supported gave action to the walking beam, a heavy timber twenty feet long. One end of the beam connected to the rod and piston of the pump; the other was free to act as a counterpoise and furnished movement. As the beam moved, the piston stroke produced a suction thrust of seven feet, sufficient to draw water up the shaft through the pipes and discharge it. The vertical wooden shaft was turned by a horse walking in a circle and hitched to a horizontal sweep. Steam power was gradually substituted for horse power in operating the walking beam pump. Several local miners developed steam pumps of their own and enjoyed considerable wealth from patent royalties. Joplin and Webb City were headquarters for foundries and machine shops producing on these local patents. The most famous locally developed steam pumps were the Whitley and Grubb models.[37]

The Cameron Pump, a portable steam model, was used too, but it had limitations since the pump had to be placed in the sump close to the water. Steam was transported to this pump by iron pipe. If the mine was over forty feet deep, miners found that the steam condensed in the line and the pump fouled.[38]

Every mine was equipped with a sludge pond into which mine water was pumped. Considerable water was required to wash and mill the ore, and the sludge pond served as a reservoir for this purpose. After the sludge pond had been filled, the mine water pumped from the shaft was channeled into nearby creeks. This practice frequently caused pollution and discolored creek waters because of the iron oxide and acid present in the water.[39]

After the miner had driven the shaft to the ore face and had taken steps to control the water, he was ready to start drifting. This was done by excavating a tunnel at right angles to the shaft to follow the horizontal ore runs.[40] The early drift dimensions were six by seven feet. The floor of the drift was gradually raised as it advanced into the ore body so as to produce a natural fall to the shaft. Thereby, any water would flow to the sump in the shaft bottom and not back up into the working face. Also, by this sloped floor, ore cans could be maneuvered more readily to the shaft opening for hoisting to the surface. Drifting was in itself a prospecting operation to determine the extent of the ore body.[41]

Beyond the shaft access to the drift, the drift dimensions were highly irregular, varying with the size and shape of the ore body. The miners adhered to Haworth's rule of "follow the ore." If the run was narrow and shallow, the drift might be practically uniform, six by seven feet from the shaft to the end of the deposit. If the deposit flared, the drift flared too, and many drifts took on the proportions of huge subterranean galleries or man-made caverns. Some had a height of fifty feet and a width of two or three hundred feet.[42]

Lateral diffusions of ore produced more of these corridors and galleries until some mine underground workings took on the appearance of a vast labyrinth or maze. Often these occurred at different levels too, for as the shaft was deepened and runs tapped at lower levels, the drift and gallery operation was repeated. These wide and high corridors required support to prevent roof caving. In stable ground, natural pillars were left about every fourteen to twenty feet. In sheet ground, the soft cotton rock character required artificial supports consisting of sills, posts, and caps of eight-by-eight-inch timbering. The sills were held in place by sharpened stakes called spiles.[43] Miners robbed the natural pillars of their ore after exhausting the mineral in the run.[44]

The breast-stope method was used to excavate the drifts and cut the underground galleries to these great heights. By this practice, successive layers of ore-bearing rock were cut away by starting at the top and working to the bottom, the miner cutting holes and blasting, much as he did in sinking the shaft. The underground worker was expected to keep a steady stream of ore cans flowing from the drift. If he became short on "dirt," as the ore-bearing rock in the mineral face or stope was called after it had been blasted loose, he and his partner spudded powder holes and detonated a new ore face.[45]

The blasting process tore loose tons of rock impregnated with ore which fell to the drift floor. Beforehand the miner had placed heavy oak planks together, making a crude platform. The ore tumbled down on it and shoveling was made more efficient. All shoveling was by hand, the miner using a large scoop shovel. There was little or no division of labor. The miner was often the part owner, operator, hoister, driller, powderman, shoveler, and mill operator, and very early even the smelterman. In later times all rock and soil, whether ore-bearing or not, was lifted out of the ore galleries and drifts and hoisted and dumped on top in the interest of efficient practice. But the early miner was more of a gouger, getting out the ore as quickly as possible; fearful that the mine might play out at any time, he gave little attention to the future. Broadhead aptly described this careless tendency and its geological results:

> As the miners have an interest to hoist as little useless material as possible, they often fill up old drifts with such material which is mostly clay, chert, and limestone. In the course of years, these masses, as the miners say "grow together," and we were astonished in visiting an old mine in Brock Hollow, which had not been worked for fifteen years, to find some old posts of timber buried in a hard conglomerate of chert and hardened clay, the whole having the appearance of an original rock.[46]

Ore blasted from the stope was shoveled into ore cans. These were of various sizes, shapes, and capacities. Wooden barrels, cut in two and fitted with a bail, sufficed.[47] As long as the cans were handled by hand topside, they could not contain material exceeding 150 pounds. Too large a can was impractical underground, too, because of haulage problems. A heavy, oak-plank runway placed on the drift floor, taking advantage of the drift's gravity fall to the shaft access, was kept wet with water. This enabled the miner to slide the heavy cans to the shaft mouth. Gradually the plank runway was replaced with wooden rails, generally two feet apart. These tracks were fitted with small four-wheeled, flat-bedded cars. Ore cans were placed on these, loaded, and pushed by the miner to the shaft access. By 1900 the wooden rails were replaced by steel track.[48]

The hoisterman stationed at the top dropped a rope fitted with a hook, and to it the ore cans, one by one, were attached and raised to the surface. Transporting ore cars too far underground was impractical, for it consumed valuable time and cut down on the mine output. Therefore, if the deposit justified it, a second shaft was sunk as soon as the drift acquired a length of three hundred feet.[49] Several of the early mines, where operators had combined three or four mining lots, had as many as six to eight shafts, especially in the sheet ground around Webb City, where the ore extended with remarkable consistency over wide areas. George C. Swallow found in 1854 that the Moseley Mine on Shoal Creek had three shafts, one forty-eight feet deep, a second sixty-eight feet deep, and the third fifty feet deep, with four hundred feet of drifting. This was exceptional for the times.[50]

Some mines had poor air circulation in the drifts, and ventilation was necessary. Drill holes, if frequent enough and undercut by the drift, could be used. Extra shafts also helped circulate fresh air through the mine workings. If these measures failed to ventilate the mine properly, then miners installed a canvas-covered tube leading to the drifts and turned outside against the wind. If the wind was inadequate, a wooden fan turned by hand or steam forced air into the tube and through the underground workings.[51]

While some of these pioneer mining methods continued to be used, the Tri-State District was entering a new age of mining techniques by 1900. This era of quasi-modernization extended to about 1930 and can be characterized as the intermediate period. After 1930 modern techniques were adopted throughout the field.

CHAPTER VI

Tri-State Mining Methods: The Intermediate Period

The years between 1900 and 1930, the intermediate period, witnessed the gradual introduction of improved mining techniques in the Tri-State District. These thirty years also saw the entry of absentee capital, professional mining engineers, larger holdings, labor specialization, and separation of miner and operator.[1] Most of the technological improvements were applied in the interests of efficiency and increased profits. Some, however, such as ventilation of shafts and roof trimming, were adopted because state safety regulations required special protection for workers.[2] Yet the terms modernization and efficiency must be used in a relative sense when applied to the region.

H. F. Bain and C. R. Van Hise warned in 1901 that too much money continued to be lost by attempting to develop the district mines in the same fashion as those in eastern areas where the deposits were more uniform, continuous, and longer-lived. The geological character of the district, they added, required that mining continue to be done by simple methods and low-cost equipment.[3] Clarence A. Wright found the same conditions prevalent seventeen years later, and he observed that "economy in the first cost rather than a higher saving by more elaborate and efficient equipment has been the prevailing consideration."[4]

Generalization, therefore, is dangerous when describing modernization of mining methods in the intermediate period. This much can be said with certainty: Wherever a deposit was extensive and fairly consistent, the large operator using modern methods prevailed, while in the shallow pockety deposit fields, the small independent operator with relatively crude methods held forth. Also, improvement in method was contingent on such factors as power and transportation development.

The type of deposit, which varied in size from a bushel basket to a convention hall, determined whether the deposit would be exploited by hand, steam, or electrical power.[5] The Tri-State District, fortunately, was situated close to cheap fuel and power resources. The wider use of these power resources contributed materially to method improvement. The pioneer miner found that abundant timber around the mines supplied adequate fuel for smelting his lead and firing the early steam boilers. As the scope of mining operations increased, the demand for more efficient fuel and power mounted. Hand and horse-ox power were largely supplanted with steam by 1900, and soon electricity and gas combustion engines were displacing steam. The Pittsburg, Kansas, coal fields supplied all the coal required for mining and smelting needs. Larger mines adopted the practice of using a central power plant consisting of two powerful engines of 150 horsepower each and a battery of boilers. Then if one engine needed repairs, the auxiliary could carry the mine pumping, hoisting, and milling needs.[6]

In 1890 the Southwestern Power Company built a hydroelectric plant at Grand Falls on Shoal Creek. This was the beginning of Tri-State electric power. As mine power needs increased, several other generating plants were established in the region.[7] A hydroelectric plant at Lowell, Kansas, on Spring River between the Galena and Baxter Springs camps, was completed in 1905.[8] The Lowell plant was expanded by the erection of a steam generating plant at Riverton, Kansas, two miles up the river.[9] In 1913 the Ozark Power and Water Company completed its hydroelectric plant on White River in Taney County, especially constructed to produce additional power for the Tri-State mines. This new facility added thirty thousand horsepower to meet district power needs.[10] The Lowell, Riverton, Grand Falls, and White River power plants were combined into the Empire District Electric Power Company with headquarters at Joplin in 1922.[11]

It was not long before coal had a competitor in natural gas as a source of steam power. By 1905 the use of electricity and natural gas had produced a revolution in power use. According to Jesse Zook, a local mining engineer and statistician, new mines and mills were fitted completely with electricity while older installations frequently kept steam plants but changed to gas burners.[12] Natural gas was pumped from Liberty, Kansas, through a sixteen-inch pipeline sixty-five miles long. The cost to consumers was ten cents per one thousand cubic feet. The district consumed twenty million cubic feet annually, 75 per cent of which was used by the mines.[13] Brittain's mine census of 1908 disclosed that four hundred companies were using natural gas as fuel for power.[14]

By 1926, fifteen thousand tons of lead and zinc were being shipped out of the district weekly. This heavy export was possible in part because of improved transportation facilities both to markets and within the district.[15] This encouraged mining method improvement too. As long as the district ores had to be hauled to Spring River for flatboating to New Orleans or to the Osage and thence to St. Louis, development was necessarily limited. Rail transportation first entered the district when the Kansas City, Fort Scott and Gulf Railroad was built to Baxter Springs, Kansas, in 1870. This line extended to Galena in 1879.[16] In 1876 the closest railhead to Joplin was at Oronogo, seven miles distant. The Joplin-Girard Railroad was incorporated in 1875 by local businessmen and miners with $600,000 in capital stock. The line was surveyed to pass through Pittsburg and the adjacent coal fields. The last spike, a lead one, was driven at Joplin in 1877, heralding completion of this thirty-eight-mile line.[17] By 1879 the Missouri and Northwest Railroad had completed a line from Oronogo to Joplin, and in the same year the Kansas City, Fort Scott and Gulf, later the Memphis Railroad, as well as the Missouri Pacific Line, had extended to Joplin.[18]

The Missouri, Kansas and Texas Railroad entered Joplin from Parsons, Kansas, in 1901.[19] By the turn of the century the Frisco had taken over all the district lines except the Missouri Pacific and the Missouri, Kansas and Texas Line.

Spurs from these railroads connected all the mining camps, and to this day the district is a network of rail lines, many of them abandoned. While evidences of such camps as Lincolnville, St. Louis, Douthat, and Century in Oklahoma, Badger and Klondike in Kansas, and Porto Rico and Thoms Station in Missouri are virtually extinct, the weed-bound rail beds and tailing piles testify to earlier mining activity. An example of how the railroads connected the camps with main lines is found in the policy of the Memphis Railroad. It was reported in 1901 that this line had switches and lines to every mine in the Chitwood camp and stood ready to extend spurs wherever and whenever the demand required.[20]

By 1926 those mines without immediate rail connections depended on motor or horse-drawn ore trucks to transport the weekly production to nearby rail points. Ore hauling in this fashion netted the transportation companies $5,500 to $6,000 a week. The cost of transportation was paid by the smelter operators who kept ore buyers in the district to purchase directly from the bin. The trucking charge was fifty cents for short hauls with a maximum of one dollar a ton. Twelve ore-hauling companies were operating in the district in 1926, generally under contract to the smelters. Horse-drawn vehicles were more numerous than motor trucks because the roads were so bad as late as 1926 that motor vehicles bogged down.[21]

A feature of transportation that had important social ramifications was the extension of the electric railway or trolley. By 1918 the camps and important towns of the Tri-State District were interlaced with trolley lines. One could board a Southwest Missouri Electrical Railway trolley at Carthage, Missouri, in the north, travel through Carterville and Webb City into Joplin, then to Galena, Kansas, and into Oklahoma through the mining towns of Picher, Commerce, and Miami, a distance of fifty-five miles. Thus, since workers could commute to the mines, they could live in the larger communities, avoiding the limited facilities of the mining camps.[22] Unquestionably no single set of factors contributed more to method improvement and over-all development than power and transportation innovations.

Shafting, the means of gaining vertical access to the deposits, was considerably improved during the intermediate period. Hand shoveling was still employed, but the hoisting of debris was made more rapid by the use of steam and electricity. Rock barriers were removed more quickly by replacing hand drills with steam and air drills. The only way to remove the solid rock and ore was by blasting, and dynamite was more effective if inserted in openings within the rock-laden ore face. Drills produced these dynamite reservoirs. The first mechanical drill was used in 1890. It consisted of a cam-movement cutting shaft and bit, mounted on a tripod and turned by hand.[23] By 1902 steam- and air-powered drills, capable of cutting a 1¼-inch hole, were widely used. These innovations enabled a miner to drill holes three to five feet deep. Each hole was charged with about four sticks of dynamite.[24] Powder placed in these deeper holes loosened more rock and speeded shaft sinking.

The size of shafts increased from the early four by five feet to five by seven feet. Some shafts were five by ten feet and seven by twelve feet. In the case of larger dimensions, double compartments were installed, one for pumping apparatus and ventilation, the other for hoisting.[25] Three-compartment shafts were common in the low yield sheet-ground mines around Webb City, Missouri, where operators had to hoist and mill an immense quantity of ore in a given time. This was necessary to handle low-grade ores at a profit.[26] State mining laws required special ventilation shafts and workers' stairway exits, in case of disaster or breakdown of hoisting equipment.[27]

Except in cases of disaster, workers and equipment were raised and lowered in the same buckets that hoisted the ore. As in the early days, the mine was cribbed to solid rock with four-by-six-inch timbers to prevent caving.[28] Shaft sinking was seldom done by the owner-operator in the intermediate period. Specialization was evident here, too. Just as churn-drill prospecting was done by contractors, so were specialists employed to open shafts and drifts. The cost of contract shaft sinking varied according to geological conditions. Some shafts to the 250-foot level averaged $2.00 a foot where the ground was soft. Moderately hard ground cost $14.50 a foot, and solid chert or hard ground cost as high as $20.00 a foot.[29] Wright found in 1919 that the cost of shafting varied from $10.00 to $30.00 a foot on a 250-foot excavation.[30]

The Joplin camp introduced a widely heralded but short-lived shafting innovation in 1910. This was the inclined shaft, cut on a forty-five-degree angle to a vertical depth of 175 to 250 feet. The ore was hoisted in a one-and-one-half-ton self-dumping skip, which was a sort of cable-drawn ore car.[31] Six mines tried this method, but it was shortly abandoned because of frequent shaft caving and hanging-up of the skip on mine sides. These difficulties were not commonly found in the vertical shaft.

That perennial problem, mine water, was better controlled in the intermediate period. Water was a constant menace, and where ore prices forced shutdown of mines, owners continued to pump the water, for, if water was not constantly controlled, the mine shaft and drifts would fill and timbers and cribbing would rot out and machinery rust. Heavy rains added to the underground flow, frequently flooding the mines and forcing temporary abandonment.[32] Mine water with too high acid content destroyed pumping machinery, requiring installation of bronze fittings. Water high in iron sulfide discharged into streams polluted the waters, destroyed wildlife and vegetation, and produced much litigation.

Just as an abundance of water was a problem in the mines, the lack of pure water was a problem on the surface, where great quantities were required in the milling and boiler rooms. As much relatively pure mine water as possible was conduced into ponds for later use in the boiler room and mill. If the water was high in iron sulfide or sulfuric acid content, it corroded and destroyed boiler tubes and was not suitable for milling. Water discharged from the mines unfit for surface use was carried out by pipes, wooden flumes, and endless chain buckets. Some mines used tank cars at pumping stations and hauled deleterious mine water to a point where there was no danger of its returning to the mine.[33]

To furnish the relatively pure water required for boilers and milling, the larger towns extended mains to the adjacent fields for mine and mill use. The Chitwood Land and Mining Company built its own water plant to supply the Chitwood camp with pure water by forming a co-operative water company among its lessees. By erecting a pumping station on Turkey Creek, half a mile west, the company's five-inch mains furnished the field with adequate water.[34]

Several operators obtained pure water by constructing reservoirs and catching rain water. Teamsters hauled water to the camps as another expedient. Deep wells, below the mineralized zone, often furnished water pure enough for boilers and milling.[35] Just as the deposits were irregular, so was the district water level. Each camp had its own. The Oklahoma portion of the Tri-State District had the greatest water problem, and constant heavy pumping was necessary to keep the shafts and drifts clear. Also, the water was heavily charged with hydrogen sulfide gas. In new shafts and drifts the gas was so strong that miners, after working a two- or three-hour shift, had to be relieved. Workmen overcome by the hydrogen sulfide gas were rendered temporarily blind. After surface air circulated through the drifts, the gas gradually dissipated.[36]

In some camps a central shaft pumping system would drain completely as much as a forty-acre tract. In others, each mine required a pump. Underground streams, springs, and surface flood waters supplied the water source. If the ground were open and the water could circulate over a wide area, a single pumping station sufficed.[37] The miners occupying such a tract paid a fixed pumping fee to the mining land and royalty company draining the tract. If the water localized because of limestone-chert dikes, each mine operator had to deal with the water individually.[38] Even with heavy pumping and sloped drifts leading to the shaft sump, the miner often worked in seepage water that sometimes was two inches deep in the drift.[39] As a result, standard equipment with the miners was a pair of shin-length rubber boots.[40]

There was no uniform type of pump used in the intermediate period, although steam was the chief power source for pumping until after 1920. The Walking Beam and Cornish pumps continued to be used. In 1900 the powerful Worthington Steam Pump was widely adopted.[41] An advantage of the Improved Cornish Pump was that it could be lowered into the mine and operated close to the water level. As the water level descended, the pump, lowered on cables, followed.[42] The Walking Beam Pump, because of its size, had to be used on the surface. In 1918 electrically driven centrifugal pumps were adopted.[43] The steam pump was predominantly of the piston-plunger type. The electrically operated centrifugal pump was used where a large amount of water had to be raised.[44] By 1920 gasoline engine or electrically powered centrifugal pumps completely dominated the field.[45]

As in the early days, a drift or horizontal tunnel was driven from the shaft into the ore deposit. Drift dimensions ranged from six to eight feet. The same methods of excavation as in shafting were used, except drifting was easier, for it was more convenient to work horizontally.[46] As the ore body heightened and widened, the miners began stoping. By cutting a stope, miners could remove the ore in an orderly fashion, producing a huge subterranean cavern sometimes over fifty feet high. The ore face, or heading, was defined by excavating debris from around it. Cuts in stair-step fashion were made from top to bottom by blasting the ore loose. Stoping began by setting up air drills on an elevated platform along the top of the ore face. The tripod drills were powered by ninety-pound air pressure from surface compressors operated by steam, gas engines, or electricity. The drill, with a twenty-foot steel section, was fitted with detachable drill points. Each drill, manned by a driller and helper, could drill up to fifty feet of powder holes in an eight-hour shift.[47]

A heading, six to eight feet high, was produced by drilling holes eight to ten feet deep, spaced two to three feet apart and distributed over the working face from the top to the bench. The holes were drilled on a ten-degree angle to produce more pull in blasting.[48] From two to four boxes of dynamite were used to load and blast a set of heading holes. Such a charge would tear loose an average of 250 tons of ore.[49]

The heading charge, when detonated, sent tons of ore down the bench to the drift floor. Beforehand, a shoveling platform of two-by-twelve-inch oak planks, known as the cokey board, had been prepared. The shoveling platform had to be moved to keep up with the stope. That ore remaining on the bench was brunoed or pulled off with short-handled shovels.[50] After the heading ore had been shoveled and trammed out of the drift and hoisted to the top, the bench was drilled and blasted much like the heading. After removal of the bench ore, a new heading was cut and the process was repeated.[51] Sometimes the stoping became so extensive that the ore face extended one hundred feet from the roof to the floor.

The stoping process had to include roof supports to prevent

caving.[52] Supports consisted of either natural or artificial material. The most common roof support was created by excavating around a pillar. These were spaced twenty-five to sixty feet apart, depending on the character of the ground, soft roof requiring supports every twenty-five feet, hard roof requiring supports at sixty-foot intervals. Where dangerous slabs were exposed, posts were installed as props. Since caving was the most common cause of mine injury, special workmen called roof trimmers entered the stope area after a charge had been set off.[53] Working from ladders with long pry rods, these specialists cleaned all loose debris from the roof. Some mines used cribs filled with waste rock as roof supports. Reinforced concrete pillars were used in some instances. The natural pillars varied in thickness from twelve to thirty feet. An attempt was made to locate natural pillars in lean ore, but this was not always possible, and often the supports consisted of pure ore.[54]

When the deposit was worked out, the natural pillars were robbed of ore, the ones farthest from the shaft blasted first. For example, miners at the Emma Gordon Company of the Miami camp blasted all ore-bearing pillars from its mines, replacing each pillar with a crib support before proceeding.[55] If the pillars were not replaced with artificial supports, the ground often caved. This is evidenced throughout the Tri-State District by the occurrence of large, open pits filled with water. The business district of Picher, Oklahoma, has been evacuated several times because of incipient caving. Sheet-ground mines, with their soft cotton-rock roofs, continued to cause the most trouble in roof supports. Natural as well as artificial supports were employed. The artificial supports consisted of ten-inch timbers with base and cap, held in place with stakes called spiles.[56]

After the ore had been broken and the roof inspected for slabs, miners entered the stope area and shoveled the mineral into steel containers called tubs or cans. These had an average capacity of one thousand pounds.[57] Shovelers, using a number two-scoop shovel, worked in pairs. The blast generally disintegrated the ore-bearing material into pieces small enough for easy shoveling. Some oversize boulders were lifted into the can by the shoveler. Boulders too large for lifting were broken by a heavy hand sledge, or, if this failed, the large rock was "popped." Boulder popping consisted of placing a stick of dynamite under the rock and shattering it.[58] In an eight-hour shift a shoveler normally loaded twenty tons of ore and earned from six to seven dollars.[59]

The cans rode on a flat-bed car fitted to a twenty-four-inch track. This track was lightweight and could be lifted and moved in sections by hand as the stope position changed.[60] Mechanical loading was attempted through the use of steam shovels as a substitute for hand shovelers. Some operators, contemplating the installation of steam shovels, in 1909 avowed that it was "absolutely essential to increase production to keep profits up. This is in no way aimed to force a lower wage scale, but simply larger output. The supply of American shovelers is about all utilized. Some operators favor hiring Italian laborers for this work, but automatic shovels are now receiving first attention."[61] Steam shovels were used to good advantage but failed in the basic reason for their installation, that of decreasing shoveling costs.[62] Only since 1940 has a mechanical shoveler been devised that could operate cheaper than the hand shoveler.

After the ore container was loaded, the miner pushed it to a lay-by or switch and returned to the stope with an empty ore can. After several cars had collected, a special worker, the tub runner, pushed the cars to the shaft. If the cars had to be trammed over 250 feet, mules were used to pull a string of cars to the shaft. Some mines adopted electric and gasoline locomotives.[63] When mules were used, a maximum of six cars was pulled, while an electric or gasoline locomotive could carry a string of fifteen to twenty.[64] It was the general practice to sink another shaft if the deposit ran beyond 300 feet, since the expense of tramming exceeded the cost of sinking a new shaft.[65]

Once the ore cars were trammed to the shaft, they were turned over to a special workman called the tub hooker.[66] His function was to signal the hoisterman to lower the cable, attach the loaded cans one by one, and give the signal to raise. The thousand-pound cans were lifted, dumped automatically into the ore hopper by the hoisterman, and returned in twenty-eight seconds.[67] There is a record of nine hundred cans hoisted in an eight-hour shift, and eight hundred cans in an eight-hour shift was considered routine.[68] Both steam and electrical hoists were used, the electrical system coming to dominate by 1930.[69]

A development in the district which came quite by accident was open-pit mining. In 1902 a cave-in at Galena was turned from an apparent disaster into a profitable venture. The owners built roads into the pit, installed steam shovels, and resumed mining.[70] A cave-in at the Lehigh camp, ten miles north of Joplin, was exploited in the same fashion five years later.[71] An open-cut mine was operated where the ground caved in the Peacock Valley camp near Galena in 1910.[72] Open-pit mining, however, remained only a secondary approach to mining Tri-State ores once a rich deposit had caved. Shaft and drift exploitation continued to dominate.

CHAPTER VII

Tri-State Mining Methods: The Modern Period

The successful Tri-State mine operator in the modern era from 1930 to 1955, had to consider a greater variety of technical and nontechnical factors than ever before as a preliminary to opening a mine. Some of these were local, though many were national and international in character. Local factors to be considered included the fact that Tri-State ore reserves after 1930 were becoming increasingly marginal.[1] Beyond the Tri-State periphery, a mine operator had to estimate the impact of new, richer lead and zinc fields in the western United States. Added competition of foreign mining operations, stimulated by the United States government through its rehabilitation and development programs, as well as the Federal Base Metals stock-piling campaign from foreign sources, also had to be considered. Because his field was becoming increasingly marginal, the Tri-State operator had to know what aid he could expect from the federal government in the form of subsidies and premium payments to keep producing vital materials.

So, of necessity, the modern era was characterized by careful estimates and planning before opening a mine. Rarely did a wildcat shaft go down after 1930. Rather, an ore body's estimated tonnage in concentrates was computed beforehand.[2] This helped to determine the extent of operations and type of equipment to be installed. Mine experts could determine, within a satisfactory limit, the tonnage output of an ore body as defined by churn-drill prospecting and the assay content of the cuttings. Also to be considered was the productive life of the mine and the assumed market value of lead and zinc concentrates. From gross value the operator had to deduct mining and milling expenses and royalties to determine the net value of the mine.

Further efficiency was produced by keeping careful records once the mine got under way. The Eagle Picher Mining and Smelting Company, the district's largest producer in the modern period, required detailed prospecting records as well as data on mechanical loaders, air drills, manpower output, and tonnage for each shift. This company also made up a weekly organization report showing the average number of men by occupation, tonnage hoisted, and tons from each man and drill shift. In addition, company executives were furnished a monthly report showing total tonnage of concentrates produced by each man shift and the amount of powder used for each ton broken.[3]

Specialization of worker function made possible another type of mine accounting—the time study. As long as the worker was driller, powderman, shoveler, hoisterman, and blacksmith, no such efficiency analysis was practical. As each step in the mining process became highly specialized, time studies concerned with mechanical and worker efficiency were possible. J. D. Forrester and F. A. Taylor reported the limited application of time studies in 1945 and predicted more widespread use of them in the future.[4]

The local, national, and international factors confronting the operator also produced a revolution in mining techniques.[5] Much of the equipment used in the modern era was adopted to promote worker safety. For example, underground lighting was introduced. Early Tri-State miners lighted their way underground with candles fastened to their caps with balls of clay.[6] Small lard-oil lamps had replaced candles by 1900,[7] and carbide mining lamps came in for wide use ten years later. By 1940, electric cap lamps, powered by small batteries, had been universally adopted.[8]

In recent times mine drifts and galleries were illuminated with electric lights much like a city's streets. To prevent dangerous shadows, many mines used 250-watt flood lamps in appropriate positions.[9] Also, roof-caving tests were conducted by the United States Bureau of Mines through the use of the geophone. The wires of a transmitting instrument were inserted in drill holes, thereby recording noises produced by rock movement. Geophone tests increased worker safety and also provided information indicating the advisability of removing pillars and proceeding with drifts and stoping.[10]

Drills were equipped with water lines that kept the drill hole wet, thus preventing rock dust, so productive of silicosis and miner's consumption, from circulating. Special devices computed the dust content of mine air after blasting. Workers employed in positions producing rock fragments were supplied with goggles.[11] The larger companies employed full-time safety engineers to make regular and systematic surveys of all mine workings and equipment. These specialists, who had authority to enforce mine safety regulations, held weekly meetings with workers to promote safety.[12]

Basically the same methods of prospecting, shafting, and drifting with stope method were used in the modern era. However, machines were substituted for hand labor wherever possible. Shaft dimensions averaged from five by seven feet to six by six feet inside dimensions.[13] Hoisting was done on electrically driven drums. The ore can, when used, had a capacity of 1,400 pounds, nearly twice that of fifty years ago. No substitute was found for dynamite in blasting loose the hard ore faces.[14]

Air-driven drills prepared powder holes for stoping. An innovation that speeded drilling of powder holes was the Jumbo Mobile Drill. This device, mounted on a caterpillar chassis, had an adjustable frame for the air drill.[15] Much valuable time was saved because it was no longer necessary to build scaffolding for high bench drilling or to move by hand the heavy mountings and drill steel. The Jumbo Mobile Drill was driven to a stope area, its boom raised to the desired level, and drilling began.[16]

Hand loading continued to be used in some mines, but it was gradually displaced by mechanical loading. Labor shortages and rising labor costs coupled with the development of improved loading equipment were chiefly responsible for replacing the muckers.[17] Power shovels, used in several mines, were of the boom and dipper revolving type, with a capacity of ten cubic feet. The mobile caterpillar slusher equipped with a forty-eight-inch blade scraped and lifted ore from the drift floor into ore cans or rubber-tired dump trucks.[18] It is estimated that mechanization, when practical, accomplished the work of hand labor in the ratio of one-to-three. Three men operated the slusher-loader. By this estimate it did the work of nine shovelers[19] and at a smaller cost per ton. Mechanical loading in 1942 cost twenty-six cents a ton compared to thirty-nine cents a ton for hand shoveling. At the same time, 60 per cent of the district's tonnage was mechanically loaded.[20]

Operators found that hoisting tonnage was increased by the use of the skip pocket. This consisted of a forty-five-degree chute which fed from the gallery floor into the hoist shaft. The chute was covered by a heavy screen of iron rods, six inches apart. Ore was trammed to the skip pocket, dumped into the chute, and fed into the skip hoist, which was raised to the top and dumped into the mill hopper.[21] Those pieces of rock and ore too large to pass through the skip screen were broken by workers called screen apes, equipped with heavy hammers and stationed over the skip screen.[22] While greater tonnage was achieved by the skip hoist, it cost more than can hoisting. The average cost of skip hoisting was fourteen cents a ton while can hoisting was one cent cheaper. This, plus capital investment required for the skip hoist, favored can hoisting.[23]

Mule and electric locomotive haulage were still used in some Tri-State mines in this period, but this type of underground transportation required trackage for the ore cars. Therefore, the drift floor had to be worked from below by another expensive drift. Many mines adopted diesel-powered ore trucks with ten-ton trailers to expedite exploitation. Thereby, graveled roads replaced tracks. The drift floor need not be level for these trucks, and deposit dips could be cut out. Another device that enabled the operator to angle into the drift floor was rope haulage.[24] A cable and winch system running from the shaft access through the drift pulled ore cars to the shaft for hoisting.[25]

Groundwater remained a problem in the district. While fluctuations in the price of lead and zinc ores forced the mines to cease operations temporarily, pumping had to continue. The district water table in the modern era was at a level that could be controlled by consistent pumping.[26] If pumping ceased, the mine drifts and shafts filled, rotting the timbers and rusting underground equipment. A sample of specialization was found in the method used by the Eagle Picher Company to combat water. With sixty mines in the district, this enterprise maintained a permanent pumping crew, equipped with heavy truck, boom, and winch to install, pull, and repair pumps in company holdings.[27]

A visit in 1950 to the Eagle Picher West Side Mine, near Treece, Kansas, revealed the ideal in modern mining procedure. The ore here was mined at the 428-foot level. A huge mined-out gallery near the shaft, with a 125-foot ceiling, contained the tool shed and blacksmith shop and made the mine a self-contained unit underground. Nearby, but at a safe distance, was the powder magazine. Fuel for the vehicles was stored in tanks near the tool shed. The preceding night, a huge bench had been blasted out of a stope in the west drift, and workmen were busy slusher-loading the ore into diesel-powered ten-ton trailer trucks. Oxygen tanks were fitted to the motor manifolds, thereby minimizing the carbon monoxide fumes. These vehicles coursed along two-lane graveled roads leading to the skip pocket. The ore was dumped on the skip screen, fell into the hopper below, and was fed to the side into a dump shaft. The dump shaft contained two square buckets, called twin skips, which were filled alternately by two men, operating hoppers fed by the screen above. The skips carried two tons each. Mammoth electric hoists, operating on a huge drum, pulled the skips up the shaft and dumped the contents into an elevated hopper with a 780-ton capacity. Standard railroad tracks ran under the giant container, and seventy-ton-capacity railroad cars were waiting to be fed from the hopper. This railroad, owned by Eagle Picher, carried the ore to the central mill at Commerce, Oklahoma, ten miles distant. A locomotive for the ore train was rented from the Frisco Railroad.[28]

A widely publicized development in modern Tri-State mining was the strip-pit operation, used as a substitute for shafting and drifting. Recently, journalists inaccurately described strip pitting as a new development.[29] Actually this method was used in the district as early as 1902.[30] On several subsequent occasions, where the ground caved on rich deposits, mining operators used this method to continue exploitation.[31] Strip-pit operations were used briefly in 1917 at the famous Oronogo Circle Mine when two acres of workings caved on this heavy producer. Soon this pit had a diameter of eight hundred feet. Stripping resumed there in a desultory fashion in 1936.[32] Modern strip pitting was inaugurated in the Thoms Station area near Joplin in 1949, when power shovels were used to open up a deposit at the fifteen-foot level.[33] In 1950 a strip pit, with a six-hundred-foot diameter, was opened nine miles north of Joplin. The operation was completely mechanized, with carryalls and bulldozers removing 350,000 tons of overburden. The sides were spiraled to enable trucks to move out with ore. The Sucker Flats section near Webb City also was being strip pitted.[34] In 1951 the Chitwood area west of Joplin witnessed some strip pitting, too.[35]

A technical development of considerable import was the consultant facilities made available to Tri-State operators by the United States Bureau of Mines. The bureau had actively carried on explorations and made the prospect data available to operators. At the request of the operators, the bureau even prospected for individual companies. Such was the case of the O'Jack Mining Company of Joplin in 1943 when the bureau allotted funds and paid for the drilling of twenty-two exploratory holes. In addition, this government agency assigned a bureau mining engineer to the company as a consultant.[36] The Phelps Mining Company received the same sort of assistance from the bureau in 1943.[37]

In addition to assistance to individual operators, the Bureau of Mines carried on a variety of programs to assist the district as a whole. These services included health and sanitation programs in the mines, studies of improved methods and safety, and extensive investigations of ore reserves.[38]

No analysis of modern mining in the Tri-State District would be complete without some comment on the exceptions. From a study of the mass of technical literature on the subject of district mining developments, one receives the impression that the only mining in the modern period was carried on by the large company, fully mechanized and thoroughly specialized. Many field trips and interviews have piled up much evidence to the contrary. Actually the facts pointed to three different classes of mining operators active in the district at mid-century. The first, and dominant, operator class was the large company. This group produced three-fourths of the district lead and zinc concentrates and best exemplifies the foregoing description of modernization and specialization. The second operator class in the Tri-State District was the gouger. Generally, four to six miners formed a sort of association or company. These gougers leased erstwhile heavy producers, largely mined out and abandoned by the larger companies because the remaining ore pockets were not extensive enough to justify further large-scale mining. By gouging out the remaining ore, this small company of worker-operators generally took out sufficient ore to make a living. And there was always the possibility they would find a rich lode which their predecessor had overlooked.

One mine near Galena, Kansas, formerly a heavy producer, was operated by gouging in 1950. These worker-operators had obtained a lease from the landowner. Water was controlled by an electric centrifugal pump and ore was hoisted in one-thousand-pound buckets by an electric hoist. The working crew consisted of six miners who formed an association and shared the work and profits after expenses for powder, tools, and electric power. A single mule was used for underground haulage, and an electric compressor furnished power for drilling powder holes. The ore was hand shoveled into ore cans, placed on cars fitted to drift trackage. As in the pioneering era of Tri-State mining, every miner-operator of this gouger association was prospector, driller, powderman, shoveler, hoisterman, and millman.[39]

The third operator class found in the Tri-State District in the modern period was the small mining company consisting of from two to six associates. This type differed from the gouger association in that the gouger worked developed mines of proven production but abandoned because reserves were insufficient to produce tonnage required to pay operating costs of mechanized operations. The small operator was unique in that he opened a completely new mine, from the surface down, in a field of small pocket deposits, avoided by the large companies because the yield was not heavy enough to justify mining on a large scale. The small operator association applied electric and gasoline motors for compressing, pumping, and hoisting. This type of operator thrived in shallow mineral ground, and seldom did his shafting extend beyond a depth of one hundred feet. No worker-member was a specialist. Each had to be versatile. As little waste rock as possible was removed, and mining at this level was far from systematic, since the size of the ore deposit was small. Ore was picked out of the debris and hoisted. Waste rock was left in the mine. Workers grubbed out the ore, working their way through the deposit, cutting only what rock was absolutely necessary to squeeze their bodies through.[40]

This class of operator best exemplified the spirit of the Tri-State District in that its successes kept alive the region's dynamic tradition, which held that Joplin, Galena, or Picher was still a "poor man's camp."

CHAPTER VIII

Tri-State Milling

After Tri-State lead and zinc ores were excavated and hoisted to the mine surface, they had to undergo the milling process to prepare them for the market and smelting. These minerals as they came from the ground were called crude ores. After milling, lead and zinc ores were designated as concentrates. The market quotations were expressed in terms of concentrate prices and not crude ore. Lead and zinc crude ore had to be milled in order to reduce the mineral to a uniform size, purge soil and other impurities, and extract the ore from associated rock.[1] Deeper mining produced crude mineral intimately associated with flint and limestone rock, and marketable concentrates had to be free of all foreign matter. Lead and zinc at deeper levels also were associated, and milling separated the two minerals. By 1908, free ore, that is, mineral unattached to foreign substances, was a rarity. Sixty per cent of the zinc and 90 per cent of the lead were actually contained in rock.[2]

Like Tri-State mining, milling underwent considerable evolution. Until 1874 only lead was sought, zinc blende being discarded because of low market prices and transportation problems. Lead mills easily, and, since it predominated in the upper strata in this era of shallow mining, few milling problems arose. The earliest miner, seeking lead for bullets or frontier trading goods, simply extracted the free ore from shallow, grass-root deposits and smelted it on the spot over a wood fire. The only milling required was washing the soil from the raw mineral before smelting. Any ore attached to rock was left.

By 1850 milling had become less elementary. Pre–Civil War miners extracted and hoisted the ore until a quantity had accumulated. Then the underground worker and hoisterman worked together to clean it.[3] They accomplished this by each using a tool called the pickawee, a small hand pick somewhat like the modern geologist's hammer. The free ore chunks were thrown into the ore bin. All mineral attached to rock was hammered loose. Mineralized rock which did not respond to treatment by the pickawee was thrown on the mine dump pile.[4]

The resumption of milling after the Civil War witnessed considerable evolution in Tri-State milling practice. Garland C. Broadhead found in 1873 that some ore was hand separated with a hammer and chisel. Much ore intimately associated with rock was hand crushed and washed in an inclined wooden sluice. While a strong stream of water played over the mineral, a workman pulled the mineralized material against the current with a shovel. The lighter rock washed out, leaving an accumulation of relatively pure mineral. This early sluice was a trough ten feet long, two feet wide, and six inches deep. Great quantities of water were used in this primitive separator, and keeping an adequate supply on hand was a perennial problem. During the summer of 1872 it was so dry in Joplin that ore mined there had to be hauled two miles to Turkey Creek for sluicing.[5]

Broadhead noted that some miners separated ore from rock by placing the material on a small wood fire. The flame was kept low enough not to roast or smelt the lead and just hot enough to crack the rock and thus free the mineral. He added that "two men or three boys can wash per day from five to ten wagon loads of wash dirt [impure ore]. Each load weighs from 1,800 to 3,000 pounds . . . and yields from 200 to 600 pounds of clean galena."[6]

By 1876 milling consisted of hoisting the ore from the shaft, skidding the ore-laden bucket to a flat rock elevated about two and one-half feet high, and dumping the contents on the rock. The limestone-chert-impregnated ore was crushed with a hand-wielded buck hammer.[7] Miners sized the ore by placing the bucket contents on a screen called a grizzly. The fine material passing through the openings was washed and binned. The coarse material remaining on the screen was sorted by hand. Barren rock was discarded,[8] while the ore-bearing rock was "bucked up" by hand power on the flat rock, then sluiced.[9]

As the mine output increased, more efficient milling facilities had to come. The extent of underground operations depended on the quantity of ore that could be processed and made marketable. The hand jig was an innovation that enabled the miners to handle more ore. Arthur Winslow reported limited jigging practiced by 1870 around Joplin. Jigging did not become widespread, however, until 1880.[10] The hand jig consisted of a water-tight wooden tank, six feet square, three feet high, and open at the top. An upright on each side was notched to support and suspend the ore screen. A slightly smaller wooden box, the ore screen, ten inches deep and bottomed with a heavy iron screen of one-half-inch mesh, was fitted into the larger water-filled compartment.[11] Iron rod supports, fitted to the tank's upright sides, suspended the ore screen and gave it free movement. A jig pole sixteen feet long, was attached to the ore screen mounting. This lever, given an up and down shaking motion in the water tank, caused the ore in the screen box to settle to the bottom. The lighter rock washed to the top and was skimmed off. After shoveling the concentrated ore into the bin, the operator placed a new load of crude ore into the screen.[12]

The jigging operation was based on the principle that zinc blende was one and one-half times heavier than the flint gravel and lead was three times heavier than the flint.[13] In the early days of Tri-State mining, however, only lead concentrates were saved. An experienced jig man could handle from six to ten tons during a twelve-hour shift.[14] Shaner's section on milling describes how the millman sprang up and down, giving short jerks to the jig pole to concentrate the ore. He recounts that a Swede, hired as a jig man, when queried after a day's work, "How do you like jiggin'?" replied, "The jiggin' is all right, but the yumpin' is hell."[15]

Hand jigs continued to be used in an age of more modern milling to prove and develop a deposit. If sufficient ore was found, then the hand apparatus was discarded and a mill with modern power machinery was established.[16] The early miner seldom held his milled ore for more than a week. Saturday was turn-in day when the mineral was sold.[17] Before 1900, Joplin was a "poor man's camp," and most miners were operating on a shoestring. Few had enough capital to carry on operations for over a week without a turn-in. Besides paying royalty fees, they had to purchase dynamite and tools and pay for food and shelter for themselves and, in many cases, their families. The abundant entertainment furnished by the theaters, saloons, gambling establishments, and other media for which Joplin was famous, even notorious, sapped any ready cash the miners might have on hand.

Before 1875 the mining land and royalty companies holding the mining leases purchased the milled ore.[18] Some enterprises, such as the Granby Mining and Smelting Company, stipulated in the leases that the miners must sell their ore to the company. Several companies bought the raw ore and milled it themselves before smelting. Such was the Riggins-Chapman Smelting Company of Joplin. This concern, inaugurating a more efficient milling practice in 1874, took all the raw ore miners turned in and crushed the material on heavy, steam-driven rollers. Next the ore was mechanically transported to the jig room where concentrating took place in two large jigs. These were operated by a steam-driven, horizontal piston movement rather than the overhead vertical hand movement.[19]

Just as steam was substituted more and more for hand labor in mining, so it was in milling. In 1881 the Empire Mining and Smelting Company of Galena, Kansas, installed a mill to accommodate its mines on Short Creek. The crushing and cleaning were accomplished wholly by steam power. When completed, the mill had a capacity of fifty tons in twelve hours.[20] This increased capacity and mechanization of Tri-State mines and mills developed in response to a need. The shallow free lead was replaced at lower levels by lead impregnated rock and lead associated with zinc blende, also associated with rock. A more efficient method of crushing and separating the ore had to be devised. Also, after 1874 a market for zinc developed.

All ore-bearing material was raised with the advent of the power mill. Powerful steam-driven crushers ground the material into a fine gravel.[21] Then the minerals were conduced to jigs where the lead and zinc were separated from the crushed rock. Jig agitation caused the lighter rock particles to go to the top of the jig table. This gravel was skimmed off and thrown aside.[22] The capacity of these early mills was gradually increased until, by 1902, some installations could handle one hundred tons on a twelve-hour shift. In the Galena, Kansas, district alone there were over 130 steam plants.[23]

This early milling allowed much of the fine ore to go out on the tailing heap. By 1900, milling companies had adopted rollers to grind the tailings to a finer size in order to increase the extraction rate.[24] When Broadhead ran an analysis of the mill discharge on the refuse heaps in 1874, he found lead loss amounted to 5 per cent and metallic zinc to 36 per cent.[25] Finer grinding increased recovery, but the loss was still great enough to make milling of tailings profitable at a later date.

Custom and central milling are regarded as modern milling innovations. Actually both were practiced early on a limited basis in the Tri-State District. Mining land and royalty companies operated smelters in conjunction with concentrating mills to insure a steady flow of uniform concentrates for their furnaces. Lessee miners often brought their crude ore to the company mill for processing.[26] In 1877 the Granby Mining and Smelting Company erected a central mill at Lone Elm, two miles north of Joplin, as an adjunct to the company smelter there. The mill room was seventy feet long, thirty feet wide, and sixty feet high. Crude mineral was brought in through a driveway and dumped in a storage bin. From this point it was transported on conveyor belts into a revolving cylinder that was perforated with half-inch holes. This was the forerunner of the modern trommel. The finer minerals sifted through into a trough and were carried to the jigs. The coarse material remaining in the cylinder was transported to the crushers, reduced to a fine gravel size, and then carried to the jig room, cleaned, separated, and carried to the smelting room and placed in bins.[27]

Central milling was the exception, however, and each mining tract had a mill after 1900. Some of the richer tracts had several mills. In the individual mills, ore was transported up the shaft and dumped on a screen, the boulders hand sorted by pickers or cull men.[28] The ore-bearing material was transported on an elevated tramway to the mill, while the barren rock was dumped outside.[29] These improved milling techniques enabled operators to carry on in low-bearing deposits that formerly had been abandoned.[30]

Whereas about 95 per cent of the lead present in the crude ore could be separated, James R. Finlay found in 1909 that only about 60 per cent of the zinc was extracted. This was because the specific gravity of zinc was less than that of lead, and zinc was intimately associated with small particles of flint, requiring much finer grinding than was being done by district mills.[31] A problem for metallurgists was to devise improved methods to extract more metal from the crude ore, especially zinc.

Another metallurgical problem of great concern to the

operator was the deposit ore percentage. In 1901 mills were running 4–30 per cent ore. This meant that from four to thirty tons of zinc blende were produced in mining and milling one hundred tons of crude ore. In that year, 4 per cent was the lowest figure at which the mills could profitably process ore.[32] Another way of expressing this is: in handling one hundred tons of 10 per cent ore, ninety tons of waste material were hoisted and processed. In the great sheet-ore channel of Webb City the operators were handling 3 per cent ore. But the volume, possible because of the extent of the deposits, made it pay.[33]

Out of the welter of milling experiments in the Tri-State District there had evolved by 1905 a definite system, recognized nationally as the Joplin Mill Practice.[34] Scattered deposits made central milling unprofitable. The erection cost of a Joplin Mill was low and its equipment simple.[35] Also, the mill was somewhat portable. When a deposit was exhausted, carpenters dismantled the mill, as well as the hoist and other equipment, and moved to a new location. When the Oklahoma field came in, scores of mills in the Joplin and Galena fields were dismantled and moved into the new producing area.[36]

Just as the Joplin Mill was flexible in operation, its capacity could be easily adjusted. This was based on the estimated tonnage output of a deposit. Where justified, large mills were built. The capacity of an average mill in 1890 was around 100 tons. By 1900 it had increased to 500 tons. The 500-ton mill became the dominant type in the district until the re-entry of central milling around 1930. In a few instances extensive deposits justified even larger mill capacity. In 1907 the Yellow Dog Mining Company built a 1,000-ton concentrator near Webb City, Missouri, to handle the extensive blanket deposit ore of that field.[37] The Netta Mine of Picher, Oklahoma, built a 1,500-ton mill in 1918.[38]

A Joplin Mill operated somewhat as follows: The mill building was framed from two-by-six-inch timbers and covered with one-by-twelve-inch wooden sheeting or corrugated iron. It was situated adjacent to the mine shaft. The hoisted ore fell through the grizzly and was carried by conveyor belts mill.[39] The ore fell from the hopper onto a grizzly—a screen made of heavy iron bars four inches apart. The smaller crude ore fell through the grizzly and was carried by conveyor belts to crushers and rollers that reduced the material to one-half-inch size.[40]

The material remaining on the grizzly was handled by cull men, who sorted the boulders. Those boulders containing minerals were broken by heavy hammers to allow the particles to pass through the screen. Boulders barren of minerals were dumped on the debris pile.[41] After crushing, the ore was transported automatically by an elevator with conveyor buckets to a trommel or revolving screen. The mineralized material, one-half inch and less in size passed through and was led to a rougher jig.[42] The oversize material was returned for additional crushing and then returned through the trommel to the rougher jig, assuring all material of one-half-inch size or less.[43]

The rougher jig had a series of fixed cells or compartments. Water circulated through these cells by means of a plunger, powered by steam or electricity.[44] A bed of concentrated lead ore six inches thick was formed in the first cell. The zinc blende and gravel washed into the second. The nonmineralized rock flowed off through six cells, each cell collecting a quantity of zinc blende. Each succeeding cell was situated a few inches lower, enabling the water to circulate through the jig, the plunger giving a pulsating motion to the apparatus.[45] As the crushed material passed from the sixth cell, it was carried as waste to the tailing elevator. A product of the jig process was a fine sand containing minute particles of mineral called smittem or middlings. This was trapped in a device called the hutch, then carried to a cleaner jig where the fine ore particles were separated from the sand.[46]

In spite of this elaborate process, considerable ore was lost to the tailing pile. The heaviest loss was in zinc, since most of the lead concentrated out. Some mills were losing one ton of zinc concentrates for every two tons saved. To cut down on this loss, sand jigs, operating much like roughing jigs, and Wilfrey tables were devised for treating fine material. The table was a wide apparatus, slightly inclined, with sides and periodic bars or riffles placed across the bottom.[47] The fine sand, ore, and water, making a slime, were fed into the table. The sand, being lighter, was carried along over the riffles. The heavier galena and zinc blende were caught and separated in the order of their specific gravities.[48] Lead, with a greater weight, accumulated on the riffles toward the top of the table, while the zinc blende, of lesser weight, collected toward the bottom.

The concentrates collected from jigging and tabling were placed in bins ready for the market. The mill discharge, known as tailings or chat, a whitish gravel, was conduced to the tailing elevator, built some distance from the mill and at a height of sixty to eighty feet so that a high dump could be built up.[49] Some of these tailing piles reached a height of three hundred feet, covering thirteen acres with a volume of over seven million tons of gravel.[50] Frequently a tailing pile reached such a height that a second tailing elevator had to be constructed on top of the heap to afford suitable dumping facilities.[51]

The concentrate bins were constructed adjacent to a loading dock, or railroad spur.[52] As the district's production increased, mining land and royalty companies entered the production field and ceased purchasing ore. As early as 1875, smelters were sending ore buyers into the field to buy lead and zinc concentrates. Small operators profited from this practice, for they could sell at the market price rather than the arbitrary one set by the company holding their leases. Frequently in winter the ore would freeze in the concentrate bins and have to be blasted loose with a charge of dynamite before an ore buyer could make a purchase.[53]

Another milling development of consequence in the Tri-State District was the adoption of flotation in concentrating. Flotation, as an auxiliary to jigging, helped produce a higher grade of concentrates and increased concentrate tonnage by catching the fine mineral that had escaped the other milling steps. This concentrating process was first tried in the district in 1916.[54] Mineral-laden sand, running from the jigs, was conduced to a container filled with a solution of chemical reagent. The first flotation ingredient was pine oil. Later a variety of chemicals and solutions were used to float out mineral particles. These included creslic acids, xanthates, aerofloat compounds, copper sulfate, and coal tar.[55] The fine grains of rock and mineral were converted into a mudlike slime. Next this mixture was whipped by gyrators until air bubbles formed. A reagent solution, such as pine oil or creslic acid, was added. This forced the galena grains to cling to the air bubbles and rise to the surface. Then the lead particles were scraped off.[56] Next copper sulfate, the only material effective on zinc blende, was added. The zinc clung to the air bubbles and rose to the top for scraping.[57]

By 1925 flotation had become an integral part of the Tri-State milling practice along with the jig and table.[58] Yet evidence that not all the galena and zinc blende was captured by these elaborate concentrating practices was that the tailing piles, the mill discharge, could still be treated at a profit. Milling efficiency had improved to such a degree that, while in 1919 only 60 per cent of the ore was derived, by 1937 as much as 90 per cent was derived.[59] Operators took steps to recover this lost metal by bringing the tailings back into the mill and retreating them, especially by the flotation process. Some retreatment of tailings had taken place as early as 1912 when conventional milling processes were applied.[60]

This operation was not completely successful, for only a fraction of the ore was recaptured from tailings. In 1916 flotation with pine-oil froth was used at Leadville Hollow camp, two miles northwest of Joplin. The tailings had been rerun previously by tables, and metallic content was down to 1 per cent. The Denver Mining and Flotation Company was able to pull out about 1,600 pounds of concentrates for every eighty tons of gravel processed.[61] Handling 1 per cent ore would not be feasible on crude ore coming from the mine, but by using tailings, the cost of underground operations was avoided. By 1929 retreatment of tailings was common in the Tri-State District when twenty-two tailing mills were in operation recovering more than four hundred tons of concentrates weekly.[62]

Up to this time, a sort of unwritten law allowed anyone to haul tailings gratis from the various mining leases. The sudden awareness of the value of tailings prompted mining companies to post signs forbidding trespassing.[63] At first, small mills were constructed adjacent to the chat piles, and the tailings were brought into the mill by a horse-drawn dragline scraper. Improved transport facilities in the 1930's, especially the ten-ton motor truck, helped produce central milling of tailings. These vehicles, loaded at the tailing piles by gasoline motor or electric shovels, carried the tailings to a central mill. Whenever possible, rail facilities were used to transport tailings to central milling establishments also.[64]

In 1939, 625,000 tons of tailings were re-treated, and all but 40,000 tons were carried in trucks.[65] By 1942 thirteen mills operating in the Tri-State District dealt exclusively with tailings.[66] Besides high loss through tailings, the low assay of zinc concentrates was a constant problem to operators. Lead concentrates assayed between 80 and 90 per cent galena in purity. Zinc concentrates, owing to their low specific gravity, were more difficult to separate from the elements.[67] Because of the impurities present, including up to 4 per cent iron sulfide and

1 per cent lead and small quantities of lime, the assay of zinc was held down to about 59 per cent.[68]

In the Oklahoma field especially, concentrates from some mines were down to 50 per cent because of a high amount of bitumen present. This made the concentrates sticky, and they clogged the ore beds. Some oil present floated and carried off fine zinc with it.[69] Several Oklahoma mines produced ores containing a high percentage of iron pyrites varying from 2 to 10 per cent. This separated out with the zinc, and operators selling such concentrates were penalized drastically. In 1908, when zinc concentrates were selling for thirty-five dollars a ton, Oklahoma operators were receiving only seven to twenty dollars a ton.[70]

The most recent development in Tri-State milling was the resumption of central and custom milling. Central milling involved a company's taking the crude ore from several of its leases and milling the mineral at one centralized place. In custom milling a central mill processed the crude ore of independent small operators at a fee. Both practices were used occasionally until 1912. The Granby Mining and Smelting Company, especially, applied central milling to its vast mining estate, requiring all miner lessees to have their crude ores processed at company mills in Granby, Oronogo, and Joplin. In 1911 the Playter Mining Company of Galena, Kansas, used central milling with some success and custom milling for one-half of the concentrate yield.[71]

Several factors, however, discouraged central and custom milling. The scattered nature of the deposits, coupled with the lack of effective transport facilities until the 1930's, produced the largest single obstacle to central milling. Another factor was the royalty situation. Most of the mines after 1900 were on leases of from five to forty acres, and the leaseholders feared that their royalties could not be accurately determined by central milling, while milling ore on their own leaseholds afforded better oversight.[72]

This reluctance was partially overcome in 1929 when the Commerce Mining and Royalty Company negotiated a series of new agreements with leaseholders and built the Bird Dog Concentrator, a central mill rated at one hundred tons an hour. This broke the tradition of the individual mill. Leaseholders were convinced when royalties actually increased because of heavier production of concentrates at a lower cost. The reapplication of central milling undoubtedly extended the life of the Tri-State District. By 1930 operators were down to marginal ores and newer, richer Western fields were coming in. Lower milling costs enabled the Tri-State to compete effectively. In 1932 the Eagle Picher Company constructed a central mill at Cardin, Oklahoma, with a 3,600-ton daily capacity. By 1950 the mill had a capacity of 15,000 tons daily.[73]

After 1950 fully 90 per cent of the district milling was either custom or central. Ten-ton ore trucks and railroads carried the crude ore to central mill sites from mines scattered over the district.[74] At the central mill, the ores were submitted to approximately the same processes as minerals received in the old Joplin Mill. Like other steps in Tri-State mining practice, however, milling became highly specialized and centralized.

CHAPTER IX

Tri-State Smelting

After lead and zinc ores are mined and milled, the resulting metal concentrates are refined and made ready to be manufactured into consumer products. The refinery process for lead and zinc concentrates is known as smelting. Both metals in their native state contain small amounts of other elements, especially sulfur. The smelting process provides heat that produces chemical reactions which free the lead and zinc from associate materials. While both lead and zinc smelting were carried on in the Tri-State District, only the former was important during the first fifty years of mining activity.[1]

Trappers and Indians mined and smelted surface lead ore near Oronogo as early as 1836.[2] The purity of this shallow-mined lead made smelting relatively simple, and it melted readily in wood-chip fires.[3] The earliest smelters were operated by Indians and traders seeking lead for bullets and frontier trading media. The pioneer refinery was an exceptionally crude apparatus, consisting of two flat stones placed in the form of a V and built on a hillside with the furnace hearth slanting downward toward the mouth. A small space was left at the point of the V forming a trench so that molten lead could flow down into a mold. This shallow lead mine and accompanying furnace were deserted when the user's needs were met, but they were revisited periodically much like the frontier salt deposits. These primitive lead works often were ringed with slag and ash-filled lead furnaces.[4]

The earliest type of Western smelter which operated on a commercial basis was the log furnace. According to Schoolcraft, it was peculiar to Missouri and was used as early as 1800 in the eastern Missouri lead mines. Constructed of stone with masonry-sealed joints, the log furnace consisted of walls three feet high and open in the front which created a space approximately four feet across. Built in the open, preferably on the side of a hill, this apparatus was fitted with a short stack to discharge the lead fume and wood smoke.[5]

Constructing the furnace on a hillside allowed a natural fall for the molten lead to flow into molds. The furnace was charged by placing a layer of logs on the ground base of the furnace. Pieces of lead ore the size of a man's fist, or smaller, were placed on top of the log base. Commonly a charge of lead approximated five thousand pounds. The contents were covered with kindling material and additional logs and then ignited. The stack draft was held close so as to keep the fire's temperature low and even for about twelve hours in order to draw out the sulfur present in the ore.[6] For an additional twelve hours the heat was gradually increased until the ore pile became molten and was drawn off.

During the first year or so of mining in the Joplin district, each pioneer miner had a furnace adjacent to his claim. All lead ore produced from the diggings was processed for market on the spot, for in addition to being an underground worker and millman, the miner was also a smelterman.[7] Tingle, Campbell, and other pioneer miners in the Joplin area at first used the frontier V furnace to refine their ores, but these chip-fire crudities were shortly displaced. Continually increasing production after 1848 required improved smelting practices and increased capacity. So the log furnace was in use throughout the field by 1850.[8] A preliminary to the masonry log furnace was the pit log furnace, constructed by excavating a hole in a hillside, wide and deep enough to accommodate the wood fuel and charge of lead ore, consisting of from two thousand to four thousand pounds. Some smeltermen along Joplin Creek modified the log furnace by placing alternate layers of lead and wood until the device was properly charged.[9]

Soon the log furnace was found to be completely inadequate for handling the growing production, although its charge had been increased by some miners to five thousand pounds. Also, the losses incurred from using the log furnace were heavy, since at least half of the ore went into a hard residue called slag, or passed out the chimney as lead fume.[10]

In 1850 a miner named Spurgeon built what was probably the largest log furnace in the entire area. A masonry apparatus, it was eight feet wide and ten feet long, with walls five feet high and floors of flat stone. An eye, or opening, two feet square, was left in the front of the furnace. Through this the fire was stirred with a long poker, and the molten lead was drawn. The logs, kindling, and lead, placed in layers, were fed through the top. Spurgeon caught the metal from his furnace in a common Dutch oven and "peddled the lead in the small border towns and Indian Country."[11]

At the same time, William Moseley struck rich lead ore on Shoal Creek near Neosho in Newton County. His mines, from April 1 to July 20, 1850, produced 100,000 pounds of ore. Using a log furnace, he was able to smelt 3,000 pounds a day, but was distressed at the nearly 50 per cent loss.[12] In late 1850 he erected a so-called Drummond furnace in the hope of improving his smelting technique.[13] This device, built of native stone and lined with fire clay, had a hearth or base sloped toward the lead well. The furnace dimensions were ten feet long and three to four feet wide. The lead charge of 1,500 pounds, mixed with small pieces of cordwood, was spread evenly over the hearth. In the first smelting phase, lasting from one to two hours, the temperature was kept low to produce oxidation.[14]

Thereupon the furnace heat was increased by adding fresh fuel to the hearth and increasing the draft, and metal began to flow into the lead well. Impurities in the molten ore coming to the top were skimmed off by workmen using chips of wood. Then the molten metal was cast into molds, called pigs, of eighty pounds each. This method, requiring a smelterman and helper, consumed about one and one-half cords of wood and yielded in the neighborhood of 60 per cent refined lead for each one hundred pounds of raw ore.[15]

Moseley pioneered in another smelting innovation when he added a blast furnace to his Neosho smelter in 1852.[16] This was known locally as the reverberatory furnace and was unique in that it required a blast of air for effective smelting. The reverberatory furnace consisted of a masonry or brick rectangular basin approximately five feet deep and four or five feet across. The sides and back were extended twelve inches above the basin, leaving the top and front open beyond this point. The basin contained a portion of molten lead on top of which the fuel and raw ore were fed.[17]

The molten lead base furnished additional heat and hastened reduction. Roasting, or oxidation of sulfur, and reduction, or smelting of the ore, were carried on simultaneously, thus eliminating the long roasting process required on the log and Drummond furnaces. This was accomplished by raising the furnace temperature suddenly by playing a blast of air over the contents. Refined lead flowed from the ore basin into a well outside the furnace when reduction had taken place.[18] Moseley's air furnace, with two stacks to discharge fuel smoke and lead fume, had the blast of air furnished by a set of bellows operated by water power.[19]

An air furnace was erected in Jasper County in 1851 by William Tingle. Situated in Leadville Hollow, near present Joplin, this refinery's air blast was provided by a horse hitched to a sweep pole. As the horse walked in a circle, the pole axis turned a shaft fastened to a pair of bellows, filling one air unit and forcing the other to discharge its contents in a reciprocal fashion. Fuel for the furnace was charcoal, produced in a kiln adjacent to the smelter.[20]

Shortly, smelting became a specialized function. Instead of each mine having a furnace, special smelting points developed in a given area. Some miners, better sneltermen than others, gradually emerged as specialists. And soon the mineral production of a given area was going to a specific point for refining. Moseley's smelter capacity was increased with another blast furnace to handle local ores, and in 1852 he formed the Moseley Lead Manufacturing Company, the first exclusive smelting works in the district. Devoting full time to smelting, Moseley purchased lead ores during 1852 in the following amounts: from the adjacent Newton County mines, 142,500 pounds; from Center Creek mines in Jasper County, 99,074 pounds; and from the Turkey Creek mines, Jasper County, 95,073 pounds.[21]

In 1853 a miner named Harklerode erected a smelter in Jasper County and took over some of the business that had been going to Moseley. In the first year of operation Harklerode's smelter handled 300,000 pounds of ore.[22]

An interesting facet in pre–Civil War smelting history was the experience of Thomas Livingstone.[23] In 1849 he operated a mine and log furnace on Center Creek near the present Twin Grove Community. Before long, Livingstone became so skillful that he devoted his full time to smelting. Handling ores from Center Creek, his business became so good that he installed a pair of blast furnaces.[24] A small community known as French Point grew up around this smelting establishment. The name was derived from a score of French sneltermen Livingstone had brought in from the ore works of southeast Missouri. Livingstone was a staunch Confederate supporter, and his smelting works and the mining camp of French Point were obliterated by Union troops in 1862.[25]

The most extensive and enduring smelting establishment of the pre–Civil War period was that operated by the Blow and Kennett Company at Granby.[26] In 1856 this concern negotiated a lease on a section of land owned by the Atlantic and

Pacific Railroad Company. Already a mining camp known as Granby had been established, and small quantities of lead were being produced. In the first year of operation the company erected six air furnaces.[27] The Blow-Kennett works, plus the other log furnaces in the area, shortly enabled the Granby area alone to produce over eight million pounds of refined lead annually.[28] Moseley, Livingstone, Harklerode, and Blow-Kennett bought ore from the local miners, paying from twenty to twenty-five dollars a thousand pounds. Until after 1854, when he was eclipsed by the Granby Company, Moseley's smelter was the most consistent operator and drew from the widest area of mining operations. In 1854 this Newton County refinery smelted from the Center Creek Mines, 270,000 pounds; Turkey Creek Mines, 110,000 pounds; and the Oliver's Prairie Mines furnished 90,000 pounds.[29]

While some miners continued to smelt their own ores in individual log furnaces, especially if they were out on the fringe of production where it was unprofitable to transport concentrates to the large refineries, most of the lead was smelted by special smelting companies by 1860.[30] The hearth furnace with or without blast equipment was the uniform type used. It had been improved to such a point that about 65 per cent of the lead extracted from the ore.[31]

Progressive refiners saw, however, that they must address themselves to several smelting problems before their establishments could function with complete efficiency. The lead ore of the district was relatively pure, assaying nearly 90 per cent. Yet, at the most only 65 per cent of the lead was drawn out. Much of the mineral was lost in the fume, and considerable lead went into slag, a mixture of lead, carbon, lead oxide, and other impurities. Also, some lead ore was of less assay, yielding even less than 65 per cent metal. Too, the stones used for the furnace bottom or hearth, as well as the sides, soon collected a residue of the lead oxide. This crusted into a mass that diminished the lead basin's volume and reduced the furnace's general effectiveness. Thus the hearth fire had to be extinguished after a week's steady operations, and new hearth stones installed. Lastly, the intense heat of the furnace cut the efficiency of smeltermen.[32]

Solutions to these problems had to wait until full-scale postwar mining operations got under way. The general smelting picture was improved in 1866 by the installation of the water back Scotch Hearth or Scotch Eye Furnace.[33] Seventy-five per cent of the metal was obtained from the ore by this innovation. A cast-iron frame, with hollow sides and back, plus a hearth and lead well comprised this new smelting apparatus.[34] Water circulated through the frame, reducing external heat, and forming a sort of insulated jacket. Thus workmen were more comfortable. The iron lining did not allow a crust of lead oxide to form, thereby maintaining smelting efficiency without long shutdowns and costly repairs.[35]

The Scotch Hearth was started by melting the lead basin full of pig lead. This molten base helped maintain a uniform heat and hastened the smelting of raw ore.[36] About three thousand pounds of lead ore mixed with fourteen to fifteen bushels of charcoal comprised a charge. This was spread over the lead basin and hearth. Two men, forming a furnace crew, were able to reduce the charge to refined lead in about eight hours. The slower operation of the Scotch Hearth assured more complete oxidation and thereby greater yield. And since less of the mineral volatized, the loss through smoke and fume was reduced. The air blast was not used.[37]

Improved smelting practices included the use of the slag furnace. Formerly slag had been thrown on the refuse pile. Fully 30 per cent of this residue material was lead, and the slag furnace drew it out. This was not a new development in lead smelting, for Moses Austin had used it to increase the yield of refined lead from eastern Missouri ores.[38] It was not used in southwest Missouri, however, until after 1865. In structure the slag furnace resembled the Scotch Hearth except it was smaller. The slag was crushed into pea-sized particles, washed to purge the charcoal and ash, and charged into the furnace with a mixture of lime flux and charcoal.[39] The slag furnace was able to process from four thousand to five thousand pounds of slag in twenty-four hours.[40] Not only was current slag production reprocessed, but after its value became apparent, smeltermen hauled in twenty years of slag accumulation from the refuse heaps of abandoned furnaces. The Granby Mining and Smelting Company in 1868, for example, scavenged slag from every refuse heap in Newton County.[41]

Lead fume, locally called arsenic, continued to pose a problem. Some of the furnace waste went out through the chimney with the fuel smoke. The chimney was generally eighteen inches in diameter and fifty feet high.[42] At times the lead fume became so thickly encrusted in the flue that smeltermen would shut down the furnace, climb as high as they could on the chimney, and strike the sides with a club, thereby freeing the obstruction. Then the lead fume would be shoveled into a wagon and dumped in a nearby creek or buried near the smelter.[43] Lead fume was handled carefully because smeltermen found that frequent contact with the refuse produced a sickness, locally called lead colic.

In 1876, G. T. Lewis and E. O. Bartlett, Philadelphia metallurgists, developed a process for capturing the fume and converting the residue into white lead and pigment for paints.[44] Their process was applied in the Joplin District in 1877, thereby converting another smelter waste product into an important market commodity.[45] Lead fume, formerly lost in the chimney smoke, was now carried from the furnace to a collecting room through conduits five feet in diameter. The collecting room contained several hoppers shaped like large inverted cones and connected to the conduit pipes. To the mouth of each hopper there was attached a large woolen bag.

A steam exhaust fan drew the fume from the furnaces through the conduits into the hoppers and then into the bags.

The gas escaped through the cloth pores and there remained heavy blue powder, lead sulfate.[46] The bag contents were collected and charged into a common slag furnace and heated much like raw mineral.[47] The resulting fume was returned through the conduits to the collecting room and forced through the hoppers into the woolen bags. The material collected in this step was a white lead powder. It was packed in barrels and marketed as Joplin White Lead.[48]

Another smelting problem was that of dealing with lower grade ores produced in some district mines. The Scotch Hearth method drew out only about 15 per cent of the lead from the leaner ore, and when the price for refined lead was low, it was unprofitable to handle this class of mineral. The Cupola method, another post-war smelting innovation, helped solve this problem by increasing the yield from low grade ore. Cupola smelting involved the use of two furnaces. First the ore was mixed with sand and roasted at a low temperature in a reverberatory furnace. The sand became a silica through heating and decomposed the sulfate present in the ore. The high sulfate content had been a major reason for its poor quality. Roasting oxidized the lead and caused it to agglomerate. Next the partially smelted ore was moved to the Cupola furnace. This was a blast furnace, circular in shape, forty-eight inches in diameter, and holding about forty tons of ore. From this mass, approximately eighteen tons of refined lead were produced.[49]

As lead smelting became more complex and specialized, requiring more capital to operate efficiently, the number of refining establishments decreased. The resumption of mining following the war had witnessed an increase in the number of smelters, and by 1873 there were seventeen lead refineries functioning in the Joplin District.[50] But the number had decreased to three by 1894, when there remained the Granby Mining and Smelting Company at Granby, the Picher Lead Company at Joplin, and the Case and Searge Lead Company at Grand Falls on Shoal Creek, near Joplin.[51] And, again, this consolidation of lead smelting facilities in the hands of a few operators was due in no small measure to the improvement of smelting practices.

Small operators could not compete with the newer, larger installations that drew more lead from the ore and made an additional profit from the by-products. The stage was set in 1882 for the concentration of lead smelting into the hands of a single company, the Picher Lead Company. A. E. Moffet purchased from Lewis and Bartlett the exclusive use in the United States of the process for converting lead fume into white lead. Moffet paid $500,000 for the patent and applied it to his Lone Elm Smelting Works.[52] In 1886, Moffet sold out to the Joplin Lead Company, and the following year the Picher Lead Company purchased the plant and process rights.[53] From 1886 on, the Picher Lead Company dominated the smelting field in the Tri-State District.

Fuel, too, increased smelter yield of refined lead in the Joplin area. The earliest furnaces were fired with logs and cordwood. By 1873 charcoal was widely used as a fuel for lead furnaces.[54] Charcoal was more condensed than cordwood, produced a more intense, yet controlled heat, and because of its size, mixed more easily with the raw lead to produce a better furnace charge. By 1880 additional fuel for the lead furnaces of Joplin was furnished by the nearby Kansas coal fields.[55] Like charcoal, nut-size bituminous coal mixed well with the raw ore to form a furnace charge. Too, this fuel maintained a steady heat over the hearth.[56] And after 1900 natural gas was imported into the district from the Kansas and Oklahoma fields for lead refining. Natural gas, however, was much more important as a fuel for zinc smelting than for lead.

No history of regional smelting would be complete without a summary description of how zinc metal was processed. First of all, zinc ore at deeper level mining occurred at least six times more abundantly than lead ore. In recent times galena was regarded as a by-product of mining and milling, zinc being the primary ore sought. Also, after 1873 most Tri-State zinc ores were smelted directly on the fringe of the district in the Kansas, and later Oklahoma, natural gas fields. Several Joplin District lead refiners erected zinc smelters in this peripheral area, extending the bounds of the Tri-State area in an economic sense. Furthermore, in the 1880's zinc refining was carried out on a limited basis at Joplin and Galena.

Zinc blende at first was a scourge to the miners, and deeper mining produced increased quantities of this material associated with lead which they called resin tiff or blackjack. Before 1873, when there was no market for zinc blende, many mines were abandoned because it was so abundant that it impeded the search for lead.[57]

George C. Swallow found great heaps of zinc blende thrown out on the rubbish heaps of the Center Creek mines and predicted that "the great abundance of ore, and the increased demand for zinc must make the workings of it profitable as soon as a cheap mode of transportation be secured."[58] Zinc blende was so plentiful in the mine refuse heaps of Granby that miners built a fort from it to furnish protection for women and children during Civil War raids.[59]

There were a number of obstacles to the early processing of zinc blende. The metallurgy of zinc was quite different from lead in that it did not smelt so readily. No local use could be made of smelted zinc, or spelter, as it was called, whereas lead could be readily smelted and molded into bullets, or later, into white lead. When mixed with linseed oil, the white lead, produced by the Lewis-Bartlett process, made the best house paint available. The market for lead was immediately local, national, and even international, whereas the market for zinc, until 1860, was limited to the eastern United States, especially New Jersey.

Even there its use was restricted to a few products, such as galvanizing iron ware. By 1869 a new era for the Joplin Mining District was foreshadowed by the establishment of zinc smelters in western Illinois and St. Louis, Missouri.[60] Lack of transport facilities to these points, however, continued to check the development of zinc blende mining in the Southwest until 1872, when the first shipment of zinc ore was made at two dollars a ton.[61] By 1873 the coal fields of southeastern Kansas were in production, and at that time a zinc smelter was established at Weir, Kansas.[62] Shortly, Pittsburg, Neodesha, Iola, Chanute, and Cherryvale, as well as Weir, became thriving zinc smelting towns.[63]

Zinc ore was carried in horse-drawn wagons from the Joplin mines to the Kansas zinc smelters, and a closer market served as a stimulus for greater mining activity in the area.[64] Many of these early zinc smelters were operated by Tri-State smelting companies. For example, the Granby Mining and Smelting Company operated the zinc smelter at Neodesha, Kansas.[65] Zinc smelting was more complex than lead, and much more fuel was required. On the average three and one-half tons of coal were required to smelt one ton of zinc ore.[66] This explains why it was more feasible to carry the ore to the fuel. First the zinc ore was roasted in a reverberatory furnace, much like that used in lead smelting, except that greater heat was applied. This step helped to draw out the sulfur present in the ore and produced oxidation. Next the roasted ore was placed in a battery of retorts. These were basins made of fire clay or pottery, each retort being four feet long, eight inches wide, and eight inches deep.[67]

As a general rule each zinc smelter molded and baked its own retorts, importing a special earth material from the St. Louis, Missouri District.[68] A charge for the retorts consisted of a mixture of roasted zinc and crushed coal. Each retort was filled with these ingredients and heated by a coal or gas fire beneath the retorts. The coal mixed with the zinc in the retort shortly ignited and furnished additional internal heat.[69] A furnace consisting of 128 retorts held approximately

5,500 pounds of ore.[70] The furnace heat forced the ore in the retort to volatize and collect in a container attached to the retort, where it cooled to a liquid state, was drawn out, and molded.[71] Thereupon each retort was cleaned and recharged. Zinc fumes drawn from the furnaces were valuable too, in that they were readily converted into sulfuric acid.[72] By 1882 at least one-half of the zinc was smelted either locally or in the Kansas smelting area. The remainder was shipped to La Salle and Peru, Illinois zinc smelters.[73]

But as smelting facilities in the Kansas region increased, less and less of the zinc ore was shipped to eastern smelters since Tri-State miners received the same price for selling to Kansas refiners as they received from eastern smelting establishments.[74] Finally some mining operators attempted to smelt zinc ores locally. In 1881 a zinc smelter was constructed in Joplin.[75] Then in 1893 a refinery was erected at Galena, Kansas.[76] Coal was imported from south-central Kansas. Zinc smelting did not succeed locally, however, and shortly the Joplin and Galena works were abandoned, and the ores were sent to smelters situated in the midst of the fuel source.[77]

A zinc smelter's capacity was measured in terms of the number of retorts contained in the furnaces. By 1900, gas was becoming the dominant smelting fuel for zinc, and the smelters were moving westward to the natural gas producing towns of Kansas. These included Iola, Chanute, Cherryvale, and Neodesha. While the older smelters were opened from time to time on a limited basis, they were largely abandoned.[78]

The natural gas pool of southeastern Kansas lasted between ten and twelve years. As the supply available for smelting diminished, the smelters either migrated to new natural gas fields or imported fuel through pipe lines. One of the richest sources of natural gas was developed south in Oklahoma near Bartlesville. Caney, Kansas, smelters imported natural gas from this field beginning in 1910.[79] By 1912, two zinc smelting plants had been constructed at Bartlesville and two plants at

Collinsville, Oklahoma.[80] Four years later the Eagle Picher Company established a zinc smelter at Henrietta, Oklahoma. The entry of zinc smelting into Oklahoma increased the number of retorts in the United States by nearly 23,000. The total zinc producing capacity in the United States by 1915 was nearly 100,000 retorts.[81]

In recent times lead and zinc smelting, with a few exceptions, was carried on basically as described earlier. Lead furnaces were increased in size and were designated as Jumboes. By-products from the lead hearth increased in number. For example, slag was not only rerun to draw out additional metal, but the residue from the slag furnace was treated in "wool" furnaces to produce rock wool and other home and industrial insulating materials. Zinc refining was modified somewhat by the use of the electrolytic process. This involved the use of an electric current to roast zinc ore, thereby removing impurities and leaving a pure zinc.

In the modern period of district development the Tri-State District smelted only lead, and even lead smelting was further consolidated. The Eagle Picher Mining and Smelting Company became the exclusive refiner of lead ores in the district. The Granby Mining and Smelting Company, long a leader in the district smelting, left the field during World War I. Eagle Picher operated lead smelters at Joplin and Galena, Kansas.[82]

CHAPTER X

Tri-State Land Tenure and Use Systems

Few mining areas in the world can match the Tri-State District in complexity of land tenure and use. Local geology, royalties, mining land companies, precedent, and restricted Indian lands all combined to produce a most unique system of mineral leasing. Geologically, the widely scattered nature of the mineral deposits and their relatively shallow depth discouraged mining on a large scale for the first fifty years of exploitation. The lead deposits, the only metal sought for a number of years, were rapidly exhausted. But this mineral was widely disseminated, and new pockets could easily be found, for the galena appeared literally at the grass roots.[1]

This type of mineral geology was best adapted to small operations—"two-man power mining," as Wiley Britton has characterized it.[2] Partnerships could not handle a large body of mining land. Therefore, landowners divided their properties into small tracts, two hundred feet square. These were leased to miners who excavated for ore. A portion of the mineral produced belonged to the landowner. His share was called royalty. If no minerals were found, the miner was under no obligation to the landowner.[3] The Missouri Legislature enacted a mining code in 1877 setting forth the conditions for negotiating mineral leases. By law, the person or company controlling the mining land was required to "post or hang the terms of lease in a conspicuous place." But the lease was not sealed until a copy of the stipulations had been delivered to each lessee. Should the lessee at any time break the terms agreed on, the leaseholder was entitled to take over the diggings.[4] Aside from these general requirements, the parties were free to enter any agreement mutually acceptable.

The lease contained the amount of royalty to be paid, the number of days a claim had to be worked, and the length of the lease. Royalty payments through the years have ranged from 5 to 50 per cent. As to length of working time, the tradition was that a miner had to begin operations on a leased tract within twenty-one days after signing the lease. He was expected to continue mining thereafter, and if for any reason he suspended active exploration for a period to exceed fifteen days, his lease was automatically forfeited.[5] If the leaseholder should be guilty of an infraction of terms, the miner could seek remedy through the county court for equity damages.[6]

Often there were local restrictions on the number of mining claims a miner could hold at one time. In 1872 the miners of Joplin adopted a rule that "no miner shall hold more than one lot, nor company of miners more than one lot to every able bodied man."[7] Both Kansas and Oklahoma devised mining codes after the example of Missouri. Neither, however, included a leasing section. When minerals were found in Cherokee County, Kansas, in 1876, miners from the Joplin District made the strike. Just as they applied the mining techniques of the Joplin area, so they introduced the mining land system used there. Through precedent, then, the small leasehold developed in the Galena camp. In Oklahoma there was a different situation. Its mining code was silent on the subject of leasing mining lands, not through an oversight, but because the state had no jurisdiction over these lands, since they were a part of the Quapaw Reservation, administered by the Department of the Interior through its Bureau of Indian Affairs.[8]

In recent times small leaseholds continued, especially in the more pockety deposits where the ore was not abundant enough to justify large operations. More than anything else, the small lease was responsible for preserving the tradition of the Tri-State District as being a "poor man's camp."[9] Irene

G. Stone has described how the small tract system worked in the Galena camp.

> The landowner or lessee usually surveys his tract into lots of about an acre each, and if it is in developed territory, where there is a fair certainty of finding ore, agrees to pay for lead ore taken out the sum of twenty-five dollars per thousand pounds of ore. . . . By this method thousands of men with no capital but their own energy have made fortunes, sometimes of large proportions. The landowner may not reap so great a profit, but there no strikes or other disagreeable features arising from employing help, and each man, being interested in the profits insures economical handling of the ores.[10]

Miners lacking capital, and most of the early ones did, often formed a partnership with local merchants, called paying partners. These furnished food, powder, and other needs from their stock, and the miners worked the lease and gave a share of the receipts to the merchants.[11] Before the Civil War, miners leased small mining tracts from farm owners and paid a royalty to them, generally 10 per cent of the concentrate yield of the mine.[12] Resumption of operations in the postwar period witnessed a new land tenure and use practice. Mining land and royalty companies were formed to control mineral lands. At first the tracts were leased from landowners by these new combinations.[13]

Later the bulk of mining lands were owned in fee simple. In the early days these concerns leased tracts of from forty to one hundred acres from the farm owners, generally for a period of five years, at an average royalty of 10 per cent. The land was then subdivided into two-hundred-foot-square mining plots and subleased to miners.[14] Royalty paid by the workers to these companies at first was 15 per cent. The miners furnished all equipment and powder, smelted their own ore, and sold to whomever they chose. Around 1865 this practice changed when the land and royalty companies started buying and smelting the ore. Miners were then required to sell their minerals to the company which paid the miners a certain price for the mineral, deducting, of course, the royalty fee. This left the miners about twenty-five dollars for each thousand pounds of lead concentrates. Wallace H. Witcombe avers that this was "unlike any other mining enterprise in the country."[15]

Shortly the mining land and royalty companies increased their activities by paying the miners a premium of $100 to $150 to sink a shaft and develop the deposit. The Granby Mining and Smelting Company followed this practice, and if ore were found under subsidy, the miner repaid the company. If the miner failed to strike ore, there was no obligation.[16] The subsidy system was inaugurated because smelting was profitable, and, to keep operating efficiently, a constant stream of ore had to flow from the mines to the smelters. Since the lead deposits were quickly exhausted, the life of the company depended on new discoveries. Thus the field constantly was being prospected.[17]

A modification of this was for the company also to furnish steam hoisting equipment, since this meant heavier production over the hand windlass or horse hoister. The miner lessees were customarily charged a three dollar weekly rental fee for the use of the hoisting equipment and necessary steam. Likewise, since much of the district had a heavy underground water flow that impeded mining and even forced abandonment of rich producers in some cases, the royalty companies started pumping their leases from a central point to keep their lessees busy with mining.[18] A pumping charge, generally one dollar a week, was made when this prevailed.

Several mining land and royalty companies required all lessees to mill the crude ore at a central company mill, charging the miners a fee for this service and further restricting the operational freedom of miner leaseholders. In most cases, by the time the miner had paid all the charges levied by the company, he barely had wages left, and often less if his mine were

Badger Mine, Empire City, Galena Mining Camp, Kansas. (*Courtesy Kansas State Historical Society*).

Panoramic scenes of operations in the Picher Mining Cam

lahoma. (*Courtesy Bizzell Library, University of Oklahoma*).

Picher Mining Camp (Oklahoma), the last boom town of

·State District. (*Courtesy Bizzell Library, University of Oklahoma*).

Netta Mine, Picher-Cardin Camp, Oklahor

urtesy Bizzell Library, University of Oklahoma).

Daisy D. Mine, Empire City, Galena Mining Camp, Kansas. (*Courtesy Kansas State Historical Society*).

Morning Star Mine, Empire City, Galena Mining Camp, Kansas.
(*Courtesy Kansas State Historical Society*).

Empire City, Galena Mining Camp, Kansa

ourtesy Kansas State Historical Society).

Empire City, Galena Mining Camp, Kans

urtesy Kansas State Historical Society).

Empire City, Galena Mining Camp, Kans

urtesy Kansas State Historical Society).

Treece Mining Camp, Kansas. (*Courtesy Kansas State Historical Society*).

a poor producer. By pyramiding its services, the mining land and royalty company gradually belied the "poor man's camp" idea, and most miners were less like independent operators and more like laborers working for a weekly wage. Mining land and royalty companies further extended their operations by developing, that is prospecting and proving, the mineral lands they held under lease. This meant that the royalty was increased still an additional 10 per cent to cover the cost of exploration, since proving deposits took much of the risk out of mining.[19]

The royalty situation among miners had become so explosive in Joplin by 1874 that a three-day riot took place, culminating in the burning of a smelter works on land belonging to the Picher Lead Company.[20] This demonstration undoubtedly influenced the reduction of royalties by some companies. For example, the Consolidated Mining Company reduced royalty on its Kansas City Bottom tract in Joplin to 10 per cent on zinc and 15 per cent on lead.[21]

Several major mining land and royalty companies had been established by 1873. These included the Granby Mining and Smelting Company, which held the Granby section of Atlantic and Pacific Railroad land under lease. The mining company paid 10 per cent royalty for mineral rights. In addition the Granby Company owned in fee simple five quarter sections of Newton County mining land, leased another in Oronogo, and also owned and leased two sections in the Joplin field. The Thompson and Graves Mining and Royalty Company leased a section of the Cedar Creek Diggings. The S. B. Corn Mining Company of Joplin leased three-quarters of a section in Newton County. Thurman Diggings were controlled by the Thurman Company. This enterprise also held two sections in Newton County and had leased the area to the Richards Mining Company of St. Louis. The Chapman and Riggins Land and Smelting Company owned in fee simple the Carney Diggings, comprising a section of Jasper County mining land.

The Jasper Lead and Mining Company owned a quarter section in Jasper County and leased a quarter section from the Picher Land Company. The East Joplin Mining and Smelting Company owned a half section known as the Orchard Diggings, as well as a quarter called Moon Diggings. It leased the latter to the Moffett and Sergeant Mining Company.[22]

By 1880 further consolidation had taken place when the Granby Mining and Smelting Company had control of nineteen thousand acres of mineral land in Jasper and Newton counties, either by fee simple or lease.[23] Twenty years later the mining estate of this company, while holding four thousand fewer acres than in 1880, had produced over thirty million dollars worth of lead and zinc and was continuing to mine one million dollars worth of mineral annually.[24]

The traffic in mineral leases had become such an attractive enterprise for investment by 1890 that shares in local mineral land enterprises were listed for sale on the Kansas City Mining Exchange.[25] Soon a mining exchange was established in Joplin to buy and sell mining lands and leases. A similar exchange was set up at Galena, Kansas, in 1896.[26] These exchanges and the profits made dealing in mining lands further consolidated mineral properties and leases. Eastern and foreign capital was also attracted into the district. It is significant that this new capital was invested not in actual mining enterprises but in mineral lands and leases. The larger mining land and royalty companies continued to sublease the lands to actual miners. Examples of foreign capital in district mineral lands in the 1890's included the attempt of an English syndicate to lease over 100,000 acres of local mineralized lands.[27]

Another group of English investors paid $600,000 for a portion of the Center Creek Mining Land Company holdings near Webb City in 1892.[28] Then two years later a Swedish syndicate purchased the Moon Hill Diggings situated on the fringe of Joplin.[29] While eastern Missouri and other mining fields witnessed outside investment in mining, milling, and smelting, as well as mineral land control in this period, Tri-State investors placed their money only in mining land and leases. And if a mining land company had done some actual mining before 1900, it had returned to leasing exclusively shortly thereafter. The Lead and Zinc Mine Inspector for Missouri described this trend in his *Fifteenth Annual Report*:

> The tendency in the Joplin district by those holding the fee or original leasehold is to follow the lead of those who have grown rich as owners of mineral land, and today it is an exception to find a large company operating mines on its own account, their business being confined almost exclusively to pumping water, milling the ore, and leasing the mining privileges. . . . The Inspector notes the tendency toward larger and larger companies controlling great bodies of land. In Southeast Missouri the large concerns are a necessity, because everything connected with mining there is on a large scale requiring much capital. In the Southwest the large companies, while not operating mines to any extent, yet they are gradually acquiring more land and do a leasing business almost exclusively.[30]

This land-control fever resulted in a tremendous pyramiding of mineral land leases and royalties. In some cases mineral tracts had been subleased twice, meaning that four layers of royalty had to be paid each time a "turn in" of concentrates took place. The man at the bottom of the royalty pyramid was the miner, the workman who alone was associated with the basis of wealth of the district—lead and zinc.

That the inspector's observations were valid is evidenced by the following property listings, selected from the holdings of the forty mining and royalty companies operating in the district in 1901. Some of the states represented by chartered land companies included Delaware, Maine, Rhode Island, Massachusetts, New York, Pennsylvania, New Jersey, West Virginia, Michigan, Illinois, Arizona, as well as Missouri and Kansas. The American Zinc, Lead and Smelting Company, chartered by Maine, held 4,000 acres of mineral land by lease and fee simple. A Missouri chartered enterprise, the Rex Mining and Smelting Company, owned 1,006 acres of mining land called the Thousand Acre Tract. The Missouri Lead and Zinc Company owned in fee simple 1,200 acres of land, 760 acres of it in the corporate limits of Joplin, while the Granby Mining and Smelting Company owned 15,000 acres, 988 of these in Joplin. New York chartered the New York and St. Louis Mining and Manufacturing Company, an enterprise leasing 240 acres around Joplin. A 155-acre tract near Duenweg was owned by the Boston-Duenweg Company of Rhode Island, while the Sterling Company, chartered by Arizona, controlled an 80-acre lease, and the United States Mining and Smelting Company of Delaware owned 80 acres. The Phoenix Lead and Zinc Mining and Smelting Company owned and leased a total of 680 acres.[31]

After 1900 the size of mining leases gradually increased. Improvements in mining techniques made it possible for a crew of miners to handle more than a two-hundred-foot-square mining lot. Many leased tracts were increased to five acres, and by the 1940's the forty-acre tract was a standard leasing unit.[32]

Miners who actually worked the leases collected on these tracts. Small mining camps soon grew up on company lands when workers were permitted to build homes so they would be close to their work. Granby started as a mining camp situated on lands belonging to the Granby Mining and Smelting Company. Originally the company leased the lands from the Atlantic and Pacific Railroad, but later purchased them outright. In 1950, 67 per cent of the land comprising Granby, Missouri, was owned by the Federal Mining and Smelting Company, successor to the Granby Company, and the occupants paid a lease fee each year to the company.[33]

The Federal Mining and Smelting Company, in addition to owning much of Granby, also controlled eight thousand acres of land around the town. Besides issuing mining permits, the company also leased home sites and tracts for gardening and farming ranging up to one hundred acres. City lots were leased for $3.00 to $4.00 a year. Farm acreages were leased for $.65 to $2.50 an acre a year, depending on whether the land was grazing land or creek bottom land suitable for crops.[34]

The Missouri Lead and Zinc Company, holding 1,200 acres of district mineral land, laid out mining lots in southwest Joplin in 1900, and a self-sufficient community grew up complete with a blacksmith shop, lumber yard, and store. A total of 350 miners' homes were constructed on the ground.[35] Picher, Oklahoma, was a mining town that grew up on lands controlled by the Eagle Picher Company. When its business district started caving, due to earlier mining directly beneath town streets and buildings, the company assisted businessmen and residents in moving their property to a safer location.[36]

Concentrate royalties of 25 per cent on zinc and 50 per cent on lead were common on these lands through this pyramiding of operations.[37] By 1910 royalty payments were considered working quite a hardship on the operator. The lease and royalty craze had become so pervasive that leases and subleases were even being made on the mine and mill tailing piles, which could be rerun for at least a limited profit. Here the royalty exacted ran from 30 to 50 per cent.[38] Some mining engineers with foresight were critical of the leasing and royalty system used in the Tri-State District, especially in view of the region's future. Clarence A. Wright probably made the most objective analysis of the land-use problem in his voluminous studies on district methods in 1913 and 1918. According to his earlier findings, operators sacrificed efficiency and safety of workers for the sake of production, chiefly because of the high royalties charged.[39] Five years later he added,

> Some of the operators and miners try to mine only the higher grade ore, leaving much of the lower grade ore that

by itself would be unprofitable to mine. In other words, when the royalties are excessive, it is often the practice to gouge the ore. In consequence much lean ore is left underground where it represents a loss or total waste of mineral resources.

Some landowners and lessors do not seem to realize that the high royalties obtained through the subleasing practice are detrimental to their interests, as their financial returns are less in the end, and many of the gouged workings are left in such condition as to hinder seriously further mining and prospecting should the operating company be obliged to suspend work. Another result is that drifts are often left in such condition that the lives of miners are endangered. High royalties also tend to reduce wages, because lessees are compelled to work the mine as cheaply as possible, and consequently the labor is apt to be less efficient.

On the other hand, the present system of leasing has an advantage in that the land is constantly being prospected for rich ore bodies, and when a rich showing is struck on a lease, other persons seek to obtain a lease on adjacent land in order to find the continuation of the ore body.

In view of the above conditions, a system of royalties based on the net profits, or of reduced royalties under the present system, especially in subleasing, or a combination of the two, would seem feasible, and at many mines would result in more efficient mining and milling, thus improving the general conditions and placing the district on a firmer and more reliable basis.[40]

Like so many other observers, though, Wright admitted that the system of leasing in small tracts had a further advantage in that "the district is not only alive, but to a certain extent free from labor difficulties."[41] Lane Carter praised Joplin as the "poor man's district," for "the fact that so many working men are leasing on their own account is one of the reasons for the satisfactory labor conditions of the district. It is in a measure, the profit sharing idea."[42]

Those who held the short-sighted attitude toward district mining land use and royalty problems had a spokesman in J. B. Guinn:

> In 1887–88 Joplin had a boom and many people rushed in and purchased land and leases. Some from Colorado who were going to show the farmers how to mine; there were others from Pennsylvania where they sink a shaft big as a mule; and there were those from England who brought their experts and machinery. Some of these are not there now. But farmers are still there collecting royalties and waiting for someone to "show me." Many of the local methods seem crude to those accustomed to more refined and economical appliances but the man who best adapts his methods to conditions has been the most successful in Joplin. There are those who take great pride in having the biggest, deepest, and best equipped shaft in the district. Then there are others who prefer dividends.[43]

Mine owners might avoid the heavy royalty charges by purchasing the mining land they were using, but the price was too high for most operators. Unimproved land in the district in 1867 was valued from $2 to $5 an acre, while improved land brought $5 to $25 an acre.[44] Mining developments shortly increased the price of these lands to phenomenal levels. District land where the geological signs were favorable for mineral sold for $50 to $100 an acre. Prospected but undeveloped mining land showing fair minerals brought $100 to $200 an acre. Some buyers paid from $250 to $1,000 an acre for developed land near producing mines. Such was the case for the Holden tract near Belleville, Missouri, which sold for $1,000 an acre. At times, just to obtain a lease, an exorbitant premium had to be paid. A two-hundred-foot-square plot of North Star Company mining land went for 22.5 per cent royalty, plus $48,000 as bonus, and the Victor Mining Company of Webb City paid $100,000 for a six-year lease on forty acres.[45] In 1926, John Quapaw, an Indian of the Quapaw

Tribe, received $105,000 bonus from the sale of a six-acre lease comprising a portion of his allotted lands near Picher, Oklahoma.[46]

As the lead and zinc mining industry extended into Ottawa County, Oklahoma, after 1905, the Tri-State land-use pattern was made even more complex because the mineralized area was situated on Quapaw Indian lands. Allotments of 240 acres each had been made to tribal members, but the use of these lands by non-Indians was restricted. A law of 1897 authorized the lease of Quapaw lands for farming, grazing, and mining, subject to approval and administration by the secretary of the interior.[47]

A total area of seven thousand acres of mineralized land was involved, and by 1930 this rich field was producing more than two-thirds of the Tri-State District's lead and zinc concentrates. The leases on Quapaw lands generally were approved for a period of ten years or as long as minerals were found in paying quantities.[48] As a protective measure, the Interior Department maintained, through its Bureau of Indian Affairs, an agent to supervise and protect the Indians' mining interests.[49] Often leases were auctioned by the local Indian agent, so as to obtain a bonus for the Indian leaseholder in addition to his regular royalty. Probably the biggest sum ever paid for a Quapaw tract was the aforementioned $105,000 paid by the Kansas Explorations Company in 1926 to John Quapaw for the use of six acres of his allotment.[50]

It should be noted that only one-sixth of the Quapaws held land in the mineralized area. The tribal population in 1950 was six hundred, and of these only one hundred received lead and zinc royalty payments.[51] In the past some members built up royalty credits as high as $75,000, others ranging from $35,000 to $45,000.[52] In 1923 the secretary of the interior invested these royalty credits in United States Treasury notes bearing 4.5 per cent interest and held for the Indians in trust. As the output of the region gradually decreased through the years, due to exhaustion of richer deposits and the entry into marginal ore mining, royalty payments to Quapaws diminished. The local agency at Miami, Oklahoma, collected royalties for the Indians from the mine operators and paid out the money to the Indian landowner. In some cases the agent simply paid an allowance rather than the full amount. It should be noted that the Quapaws held their allotted lands in fee simple, so royalty payments went to individual members rather than to the tribe as a whole.[53]

Additional safeguards to protect the Quapaws from exploitation included the appointment of a Bureau of Mines engineer to work in liaison with the agent. The mining engineer helped supervise the leasing as well as the prospecting, mining, and milling of mineral from Indian lands.[54] Other protective features included the fabrication of an aerial mosaic map of the Quapaw mineral producing area. The Army Air Corps in 1927 aerially photographed the region to aid in administering Quapaw mineral lands. The photographs were assembled, reduced to scale, and annotated. Thereby a complete identification could be made of all mineral leases and the stage of development determined.[55]

In spite of all these safeguards there is abundant evidence to show that the Quapaws were exploited. As early as 1907, mine operators, organized through the Baxter Springs Mining Exchange, lobbied for a relaxation of the leasing restrictions and general governmental control over Quapaw lands.[56] A. L. Morford, veteran mining writer and newspaperman, revealed that: "One of the men interested in the movement is former governor Samuel Crawford, of Kansas, who now owns farm land near Baxter Springs. The former Governor . . . has had much to do with Indian affairs and in shaping legislation for the small tribes of the Territory."[57]

None of these pressure groups were successful immediately in getting the regulations removed. But the restricted period for the Quapaw lands was to expire in 1921, and several alternative plans for protecting the Indians were advanced. One of these provided for the appointment of a guardian for each Indian allottee. Since mine operators would have to deal with the guardian instead of the Indian agent, they were opposed to the proposed change. Undoubtedly their opposition grew from the fear that a guardian would be more zealous in protecting the Indian's property rights than the Indian agent had been. In 1920 the *Engineering and Mining Journal* reported on this opposition.

> It is probable that a bill to continue the restrictions, so far as Quapaws are concerned, will be introduced in Congress with their [the operators] backing. The operators believe that it is much safer and satisfactory for them to continue to deal with the Indians themselves or their guardians appointed by the county commissioners of Ottawa County. Any special guardians would be imbued with making as good a showing as possible for their wards and might be unfair in the end to the operators.[58]

But a law of 1921 extended the jurisdiction of the Department of the Interior much as it had been before.[59] Actually the facts show that the operators were aided materially by the government, in that leases negotiated during the Harding administration were more in the interests of the operators and less for the Quapaws. Secretary of the Interior Albert Fall, notorious for Teapot Dome and other scandals, also figured prominently in some Quapaw mining land skulduggery. By the law of 1921 the secretary of the interior was given authority to remove restrictions and permit the outright sale of Quapaw lands if the owner were adjudged competent. Paul Ewert, representing the mining operators, negotiated the sale of a mineralized tract when the Department of the Interior adjudged the owner competent in 1922. Soon it was discovered that Ewert was an employee of the United States Department of Justice. Thereupon the Quapaws sought, unsuccessfully, to have the transaction set aside, claiming that Ewert was disqualified because of his official connection with the government.[60]

During the same year, Indian attorneys attempted to gain a 5 per cent royalty increase. The operators countered by seeking a reduction from 10 to 7.5 per cent. A Department of the Interior hearing was conducted at Miami after which the secretary of the interior approved the leases at 10 per cent over protests of the Quapaws.[61]

These actions raised a storm of protest among groups concerned with Indian welfare. In 1924 the Indian Rights Association attacked the way in which the Department of the Interior was handling Quapaw mineral lands, publishing an account entitled *Oklahoma's Poor Rich Indians.*[62] Two years later, ten Quapaws holding mineralized allotments filed suit in Federal District Court at Tulsa to recover thirty million dollars worth of lead and zinc concentrates taken from their lands since 1921. Their bill of complaints charged that "Former Secretary of the Interior Albert Fall executed leases over the protests of the Indians, and on a lower basis than other offers at the time, and lower than estimates made by the Department of the Interior as to what a fair price would be."[63]

Defendants included the Eagle Picher Company as well as the Hunt, Commonwealth, Kelton, and White Bird Mining companies as sublessees of Eagle Picher. A decision was handed down in 1928 in favor of the defendants and upholding former Secretary Fall's action.[64]

Another significant development during the 1930's in Tri-State land use was the commingling agreement, whereby the ore from several tracts was hoisted and milled at a central place. Even though an ore body occurred close to the tract lines of several lessees, the minerals previously had to be mined, hoisted, and milled on the tract of its origin. Therefore, each tract, whether two hundred feet square, five acres, or forty acres, had to have at least one mill. As long as the ore was rich, the system worked fairly well. But as more marginal ores were tapped, individual hoisting and milling costs forced scores of mines to close. H. A. Andrews, Indian agent at Miami, one of the authors of commingling, reported that this new land-use practice made possible a new mining era in the district. In addition to prolonging the region's productive life, Andrews added that commingling assured consistent, if somewhat lower, royalty payments to Indian as well as non-Indian leaseholders. This practice also furnished more reliable employment and enabled companies to carry on in periods of metal price depression.[65] Basically, commingling produced lower mining and milling costs, thereby enabling Tri-State operators to handle marginal ores and compete favorably with newer, richer lead and zinc fields in the West. Likewise, commingling was of major importance in producing another major development in Tri-State mining history, that of central milling.[66]

Ironically, while commingling and central milling were the answer to Tri-State operational problems, they produced a new problem for mine owners in the realm of labor relations. Historically, the region has been violently anti-labor, and, until fairly recent times, operators were able to frustrate unionization attempts. But in 1942 the key establishment in the district, Eagle Picher's Central Mill in Commerce, Oklahoma, was organized. A strike at this plant, as in the months of May to December, 1953, had the effect of closing the district. This was because most of the crude ore mined in the Tri-State District was commingled and processed at this key establishment. Thus the cause of organized labor was promoted through commingling and central milling.

Diagram A shows the system of land use in effect before commingling was adopted in the Tri-State District. Even though the same ore body might appear on several leased tracts, the earlier practice required that each tract must have an independent shaft and mill. As long as the deposits were

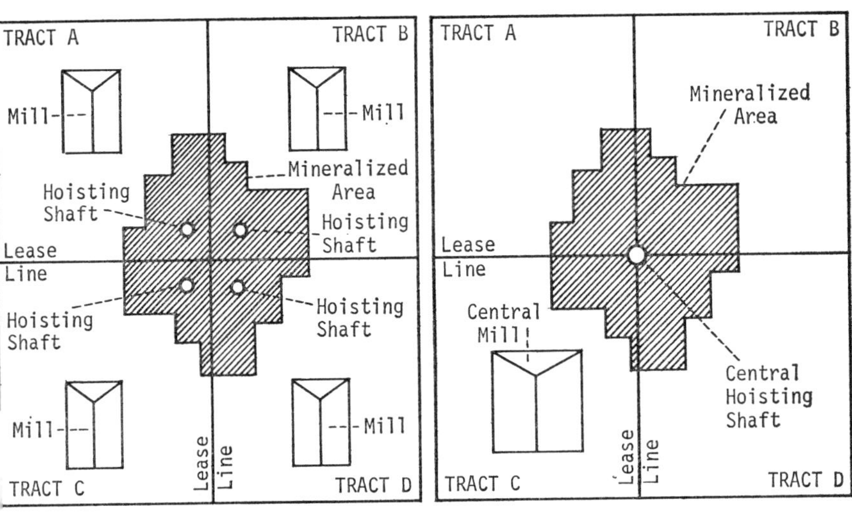

DIAGRAM A

DIAGRAM B

rich, this system did not work too badly, but as more marginal ores were reached and the richer ores became exhausted, many mining companies suspended operations because they could not handle the leaner ores profitably. Around 1930, H. A. Andrews, Indian agent at Miami, and other leaders in the field, introduced a new system of land use they called commingling, illustrated by Diagram B. By this system a single hoisting shaft could be used in common by several companies exploiting the same deposit on their respective leases. Also a central mill handled all crude ore and produced the concentrates, further cutting the cost. Land owners signed these commingling agreements permitting ore from their properties to be hoisted from a central shaft, as well as allowing the ore to be mixed and milled centrally. Where commingling involved Indian lands, it was necessary for the United States Department of the Interior to approve the agreement.

CHAPTER XI

Tri-State Mine Finance

From the standpoint of business development and organization the Tri-State District was not unlike other mining regions in the United States. Trends in company consolidation and specialization, the entry of outside capital, and the pyramiding of corporate control, common to Western mining fields, as well as in other industries, also took place in the Tri-State District. Yet, there are sufficient exceptions and anomalies to merit notice.

First of all, outside capital was attracted to the district, but before 1910 these investments were largely confined to mining land and royalty companies. Secondly, the actual operation of mines was left to small operators, who, with only a limited capital, mined the scattered lead and zinc deposits with rather primitive equipment. The general absence of capital at this level helped to produce the "poor man's camp" tradition, and in spite of the lack of the benefits of heavy capitalization, the production of ore by a single man in one day was greater among Tri-State miners than in any other mining region in the country.[1] Lastly, the marketing system used in the Tri-State District was quite different. Instead of operators shipping to smelters and paying freight charges, as was the case in the West, ore buyers came to the Tri-State District, purchased concentrates from the operators, and paid the freight charges themselves.

From 1848 to 1872, capital, the lifeblood of commercial mining, was furnished locally, either by the miners or businessmen from Granby, Carthage, or Joplin. The first mining companies were partnerships. The company members shared the work as well as the profits and losses of mining. One method used to grubstake operations was for the miner partners to take in merchants who furnished food and supplies. In return these paying partners received shares of the profits.[2] Another way mining enterprises were financed was for local doctors, lawyers, and bankers to form mining companies in partnership with experienced miners, the former furnishing the money, the latter the skill and labor.[3]

Workers themselves invested in mining operations too, but in a different fashion. They would work long enough to save a grubstake from wages, then form a partnership, secure a lease, and mine until they struck ore or until their funds ran out.[4] Small operators of this sort always needed ready cash, so it was common to sell the prospect outright once minerals were struck or to sell a half interest to businessmen and continue mining on a share basis. Such was the case of J. H. Frye and his son, who had worked as miners on a wage basis, saved their money, and used the savings to cover their prospecting near Diamond, Missouri. The Fryes struck ore within three days, and, to carry on with improved equipment, they sold one-half interest in their prospect to businessmen from Kansas City, Missouri, for $4,200 cash.[5]

These early miners paid production royalty to the land company. Also, they were compelled to sell their ore to the company before smelter consolidations took place. For any services the landowner provided, such as pumping and furnishing of hoisting equipment, the miner paid a weekly fee. As noted earlier, Wallace H. Witcombe called this system of leasing and small operations, where capital of any extent entered only the land and royalty level, "unlike any other mining enterprise in the country."[6]

In order to promote outside investment in the district, several methods were used. As noted earlier, the emphasis on investment before 1910 was on mining land and royalties rather than actual mining. District advertisements and prospectuses published before 1910 emphasize this. The earliest publication of this sort appeared in 1865. It was *Parker's Missouri Handbook—Information for Capitalists and Immigrants.* Nathan H. Parker encouraged persons of initiative and ambition to immigrate to the lead fields of southwest Missouri. Nine years later, E. Lloyd and D. Baumann published a mining prospectus entitled *Mineral Wealth of Southwest Missouri.* While this stressed the mineral production and general prosperity of the region, it emphasized investment opportunities in mining land. Probably the most extravagant description of Joplin's resources was contained in the *Klondike of Missouri,* published in 1893 for the Kansas City, Fort Scott, and Memphis Railroad. This prospectus presented accounts of successful land and royalty companies, in addition to testimonials of mine operators who had made fortunes in the district.

Abundant write-ups in the *Engineering and Mining Journal, Mining World,* and *Western Journal and Civilian* gave national as well as international publicity to the Tri-State District and the opportunities for investment and profit there. One of the most persuasive bits of literature on district opportunities was Otto Ruhl's *The Rationale of Investment in Zinc Mining,* published in 1910. Even in the official reports, investors could find admonitions and helpful information concerning the district. H. F. Bain and C. R. Van Hise, after describing the geology of the district in their *Preliminary Report,* concluded by writing a guide for persons interested in investing in Tri-State lands and mines.[7] Clarence A. Wright in 1913 and 1918 used U.S. Bureau of Mines facilities to warn investors of the problems as well as the opportunities of Tri-State mining enterprises.[8]

Besides mailing abundant literature to investors, Tri-State promoters maintained special agents through the Ore Producers' Association to attract Eastern capital. One agent, Frank Nicolson, wrote in 1899:

> The unique feature of mining here is the high percent of net profit to total value of output and large number of successful properties out of the total number promulgated. Throughout the West, around ninety percent of the mines operated with ample intelligence and capital soon pay out and fail. In this district, at least ninety percent of the mines so operated succeed. Since January 1, 1899, my clients have purchased about $2,500,000 worth of property and is paying dividends at the rate of from one percent to three percent per month. As far as I know, no similar record can be furnished by a mining district in any part of the world.[9]

A sample of boom articles appearing in the *Engineering and Mining Journal* is a local description by Lane Carter, a mining engineer from Chicago.

> As a rule, mining camps and mining towns in America are not the most desirable places to live. No one would select for instance a residential site in the coal mining districts of Wyoming, or the sagebrush regions of Nevada for his home if he could help it. But there are mining centers in the United States where one enjoys the amenities of life and gets away from that hideousness so often concomitant with mining. Such centers include . . . the zinc district of Joplin, Missouri. The visitor can see mines in every direction stretching away into Kansas toward . . . Oklahoma. But he does not see the landscape darkened by the clouds of black smoke which hang like a funeral pall over most mining cities. . . . The romances of mining are not confined to gold, silver and diamonds. The Joplin district, the largest producer of zinc in the United States has many stories of men who have struck it rich and made fortunes, and there are few places in the mining world where the chances are better, for the man with a moderate amount of capital of $25,000 may do a great deal, while with $50,000 he may make an investment for an independent life. Capital is necessary for success, but compared with the large amounts required in gold and copper mining, the sum needed in the zinc district is small.[10]

There were several reasons why investments in the Tri-State District were so attractive. The scattered nature of the deposits discouraged heavy outlays for plants and equipment, so the amount required was considerably lower than in other regions. Walter R. Ingalls noted that in all the Western districts the cost of mining and milling a ton of crude ore was considerably higher than in the Tri-State District. Joplin wages were lower and the output by each man each day was higher than anywhere in the country.[11] "Joplin is at the present about the most prosperous mining district in the country,"[12] and "stockholders draw regularly dividends of not less than 20% on their investment,"[13] became common phrases in the investment world. Under these conditions, outside capital eventually entered the field. For the most part, investors negotiated with district representatives rather than coming in person to the Tri-State District.[14]

Jay Gould visited Joplin as a guest of the Joplin Business Club in 1890 and, after being shown the opportunities for investment, wanted to know "how a district so rich in natural resources, and of so considerable development, had so long escaped his notice and that of his lieutenants."[15] John D. Rockefeller came to Joplin twice, and his visit in June, 1892, prompted the editor of the *Engineering and Mining Journal* to observe: "There are indications that lead and zinc investments of the state are about to pass into control of foreign capital. John D. Rockefeller recently looked over the ground and his plan is to obtain control of the best part of the production of the smelters and properties for a proposed trust that has as its object the controlling of prices of lead ore and jack."[16]

There is no evidence, though, to show that either Gould or Rockefeller invested in the Tri-State District. Eastern capital, however, did enter the district in the 1890's, and within twenty years many enterprises were characterized by absentee ownership. Eastern engineers came in too. For example, James Gallagher of Boston was selected to direct the Joplin Mining Stock Exchange in 1907,[17] and Frank Playter, also of Boston, became one of the most famous of the district mining engineers.

Foreign capital was also attracted to the district. In 1892 the Center Creek Mining Company sold some of its lands near Webb City to an English mining syndicate for $600,000.[18] Another English mining combine negotiated a sale of over 100,000 acres of mining land in southwest Missouri in 1890.[19] Then in 1894 a group of investors from Guttenberg, Sweden, purchased the Monkey Hill tract near Joplin.[20]

This welter of investment sources—local, national, and international, had produced at least fifty sizeable land and royalty enterprises by 1910. By far the largest was the Granby Mining and Smelting Company. This firm was organized in 1857. Eight years later the company was reorganized and its headquarters moved from Granby to St. Louis.[21] Then in 1873 the enterprise passed into Eastern hands; the president was from Connecticut, its vice-president from New York, and the secretary was from St. Louis.[22]

By 1900 the Granby Mining and Smelting Company held some twenty thousand acres. From 1865 to 1900, thirty million dollars worth of concentrates were marketed from its lands, and this industrial estate continued to produce ores valued at one million dollars annually.[23] The extent of small operations on its lands is shown by the fact that there were 1,200 producing shafts on its Granby premises by 1873.[24] Patrick Murphy, cofounder of Joplin, formed a land and royalty company in 1872. This was the West Joplin Lead and Zinc Company, responsible for developing the Sherwood, Cox, Pinkard, Leadville, Carney, Stephens, Tanyard Hollow, and Taylor diggings. Murphy also organized as a subsidiary the Empire Mining and Smelting Company, responsible for the development of the Galena field.[25]

Just as there was a combination in land control and mining,

so was there the same trend in smelting. Formerly, every mine had its own lead smelter.[26] Then for a brief period smelting was taken over by the land and royalty companies. But by 1895 there were only three smelters in the district, the Granby Mining and Smelting Company at Granby, the Picher Lead Company at Joplin, and the Case and Searge Lead Company at Grand Falls.[27] In 1950 there were only two smelters in the district, one at Joplin, and one at Galena, both owned by the Eagle Picher Lead Company.

Due to the progressive entry of outside capital and control into the district, there were, by 1900, fifty-three mining land and royalty enterprises, incorporated by thirty states with a total investment of $17,615,000 and holding 45,324 acres of mineral land.[28] This year marked a peak in the pyramiding of mining lands in the Tri-State District. The *Engineering and Mining Journal* reported in that year that the Anglo-American Mining Syndicate, capitalized at $10,000,000, was in the process of consolidating "some of the better producing properties in the district." Anglo-American's London agent, Henry Seeley, set up headquarters in Joplin and carried on negotiations for his concern. He offered a total of $3,000,000 for the Southside Tract of Galena and the Rubberneck Tract of Neck City. At the same time the Zinc-Lead Company of America secured options on $3,000,000 worth of mining properties in and around Joplin and consolidated them under one management.[29]

These land and royalty enterprises were formed to collect 20 to 50 per cent royalties on all mineral produced. This involved no expenditure for equipment and labor. Company lessees, the operating companies, took all the mining risks. For example, the Granby Mining and Smelting Company operated only a single mine in this period. Yet, while capital flowed for the most part into land and royalty enterprises, some small amounts were attracted to actual mining operations. It was quite common for absentee investors to hire a local miner to operate their mine. The Lackawanna Mining Company held a lease of three mining lots from the United Zinc Company at Chitwood. W. S. Mears, a Joplin miner, managed the mine, controlled by Scranton, Pennsylvania, businessmen.[30]

The American Zinc, Lead, and Smelting Company, controlling four thousand acres of district mining lands, was formed in 1899. During its first year this enterprise operated fourteen mines and mills near Carterville. But as a Missouri mine inspector pointed out in 1901, "like most of the other large holders of mining lands, it has changed its policy, and no longer mines ore on its own account, but does a purely leasing business."[31] Oddly, while production of concentrates was the basis of royalties which these companies collected, only limited capital was applied to mining proper.

Thus there existed the anomaly of modern high finance in an area of primitive methods, including horse hoisters and hand windlasses. The records show, however, that money could be made on purely mining investments, both by individuals and companies.[32]

The experience of Leonidas P. Cunningham, repeated by Tri-State miners time after time, illustrates how a worker could amass a large fortune through mining. As a pioneer miner at Joplin, Cunningham had accumulated enough capital by 1875 to enable him to underwrite the Joplin-Girard Railroad, a $600,000 enterprise. By the turn of the century Cunningham had entered the local banking business, where he remained until his death in 1911. The *Engineering and Mining Journal*, noting the probating of the million-dollar Cunningham estate, pointed out that, "Although Leonidas P. Cunningham made a fortune in the Joplin mines, he wills that if any of his beneficiaries invested their bequests in mining property they should forfeit their right to participate in the division of the estate."[33] Mining companies made a good record of earnings in this period too. The Webb City Mining Company, capitalized at $55,000, paid monthly dividends of

2 per cent on investments.[34] The Sultana Mining Company, with five mining lots in the city limits of Joplin, represented an investment of $8,000. A single shaft produced $34,878 in the first year of operation.[35] The I Know Mine of Leadville Hollow for the year 1895–96 paid a 300 per cent dividend on capital stock.[36]

Otto Ruhl gives the example of a mine requiring $35,000 investment, which in three years of production averaged a concentrate production of $20,000 annually.[37] By 1910 outside capital in large measure had retreated. The pyramiding of royalties to as much as four layers had reached the point of diminishing returns, in spite of the fact that mine production continued to increase. For example, lead production advanced from 29,055 tons in 1900 to 43,840 tons in 1910, and zinc production jumped from 246,087 tons in 1900 to 296,519. The combined value of this production by 1910 was $13,600,721.[38]

Yet, it was not sufficient to pay, in addition to labor and operating costs, the royalty dividends exacted at the various layers. Erasmus Haworth noted this trend as early as 1904.

> Little outside capital came into Galena until the spring of 1899. The unusually high price of ore at that time was responsible for an extensive advertisement of the whole Missouri-Kansas area throughout the East and North, resulting in a great influx of capital and promoters who would buy property here and then sell it to newly formed companies. Many properties sold for from two to five times their actual value, and a great deal of outside money was so poorly invested that it was entirely lost. These bad results were due partly to a lack of proper management by the new companies. Frequently, young men, sons or relatives of leading stockholders, would be sent to the mining region to have a general superintendency over the property bought at a large price. Too often such superintendents discharged their duties principally by hotel corridor conversations. Such inattention was more largely responsible for failure than any other cause.[39]

And the *Engineering and Mining Journal* reported in 1910 that little new capital was entering the district, for "so much has been lost by outsiders who do not know the district."[40] Ruhl's *Rationale* came out in 1910 as an appeal for reluctant capital to return to the district. Compared to the *Klondike of Missouri*, published in 1893, considerable change in tone is evident; the former was the product of a new, more conservative era; the latter mirrored a reckless, wildcat investment age that was on the way out by 1910. Ruhl issued a set of rules for investors and each statement stressed caution.

In 1913 and again in 1918, Clarence A. Wright used facilities of the United States Bureau of Mines to produce a guide to investors. He encouraged capitalists to put their money into actual mining companies rather than land royalty enterprises. This, he pointed out, would reduce mining costs as well as royalties and thereby increase the profits of mining. He advised such mining companies and corporations to know the extent of the ore deposit, calculate its life and tonnage through drill development, and gauge the mining and milling operation on the following basis. "Economy in first cost, rather than a higher saving by more efficient equipment is the ruling consideration. Increased capacity at small cost is . . . essential at mines where the ore bodies are as low grade as they are here. The small size and the low grade of most deposits must lead to conservative investments."[41]

While slipshod investment and poor management, in addition to royalty in areas of diminishing returns, ruined most of the land companies, some operating concerns were guilty of malpractices too. Often a company with at least moderate backing operated a year or more before its management was aware that a deficit existed; that is, the ore produced was not paying operating expenses. In 1911, Ruhl sought to overcome this by advocating the installation of a mining cost accounting system. He noted that out of the fourteen million dollars worth of concentrates produced in 1910, only about one-fourth was on the basis of cost accounting. He recommended charging labor, powder, equipment, management, and other costs to each ton of ore hoisted on a daily basis. In addition to advising operators if a deficit existed, Ruhl added that the United States Congress House of Representatives' Committee on Ways and Means would find such a system of cost accounting helpful when that group considered tariff boosts for the protection of the lead and zinc industry.[42]

Between 1915 and 1930 a new era of mining capitalization and operation took place. Local men largely dominated the scene. Many companies bought the mining lands which they had been leasing, and, more and more, landowners leased to operators rather than to mining land and royalty companies. This scaled the royalty from the exorbitant 30 to 50 per cent levels to between 8 and 15 per cent, with 10 per cent becoming the standard royalty charged. While actual production in the era of the mining land and royalty companies had been largely in the hands of a multitude of small operators, the new era of mine finance saw the rise of thirty large operating companies which by 1925 were producing 75 per cent of the district's output. Many of these were organized out of the remains of mining land and royalty companies.[43]

These large operating companies included the Federal Mining and Smelting Company, the Eagle Picher Lead Company, the Commerce Mining Company, the Admiralty Zinc Company, Goldenrod Mining and Smelting Company, Bethel-Domando-Croesus Company, the Acme Mining Company, Ramage Company, the Vinegar Hill Mining Company, the Montreal Company, the Hettig Company, Childers Mining Company, Wallower Mining Company, Kansas Explorations Company, Laclede Lead and Zinc Company, Commonwealth Mining Company, American Lead and Zinc Company, and the Underwriters Mining Company.[44]

There was less mortality among these larger combinations. They could operate longer at a loss and maintain operations in the face of lower ore prices common in depression periods.[45] One of the chief reasons for the rise of large operating enterprises was that much of the production after 1925 was coming from Oklahoma. In this year, Ottawa County produced 103,359 tons of lead and 549,211 tons of zinc. Southeastern Kansas produced 29,480 tons of lead and 228,383 tons of zinc. Southwestern Missouri produced 3,569 tons of lead and 28,456 tons of zinc.[46]

The Oklahoma deposits were more continuous and justified more extensive capitalization and operations. Also, the ore ran deeper, requiring improved equipment. Soon, better roads and motor trucks, plus added railroad facilities, made central milling practical. This produced additional consolidation because lead and zinc ore has no value until concentrated by milling. Since 1930 still further consolidation occurred until, in 1950, the district was dominated by one producer, Eagle Picher. More than one-half of the Tri-State District concentrates were produced from this company's mines.[47]

This giant enterprise operated thirty-eight of the important mines and milled fourteen thousand tons of ore each day. Eagle Picher started as the Picher Mining Land and Smelting Company in 1874. Before the century was out, the Picher Company had purchased the Bartlett Process for converting lead fume into white lead and was the leading smelter in the district. Its first refinery was the smelter at Joplin, and in 1912 it took over the Galena Lead Smelter. In 1933, Eagle Picher purchased zinc smelters at Van Buren, Arkansas, and Henrietta, Oklahoma.[48]

Eagle Picher operated metallic products plants at Cincinnati, Ohio, Newark, New Jersey, and Argo, Illinois. It had a research laboratory and rock wool insulation plant at Joplin. In 1942, Eagle Picher commenced mining lead and zinc near Tucson, Arizona. The ore was treated in a 150-ton flotation mill.[49] At the same time, the company started mining lead and zinc at Taxco, Mexico, with a mill to handle 700 tons of ore daily.[50] In addition, Eagle Picher carried on explorations for lead and zinc in Canada.[51]

In Tri-State District mine production, Eagle Picher became active in the Commerce and Picher fields. The company was responsible for discovering the rich deposits at Picher. With four heavy producing mines to start with, Eagle Picher took over the Consolidated Lead and Zinc Company in 1933, with seventeen mines, and the ten mines of the Mary M. Mining Company in 1936. In 1937, Eagle Picher dewatered several mines in the Webb-City–Oronogo area including the Oronogo Circle, and carried on a strip operation there. A year later the company purchased the Commerce Mining Company holdings, acquiring eighteen heavy producers.[52] At the same time the firm acquired the Mutual Mining Company mill at Oronogo to central mill peripheral ores.[53] Also in 1938 the Eagle Picher Company developed a central mill near Cardin, Oklahoma, with a capacity of six thousand tons a day. Soon this was increased to ten thousand tons a day.[54] The Cardin Mill, largest in the district, had a capacity of fourteen thousand tons a day.

Other large companies operating in the district have been the Nellie B. Mining Company, second to Eagle Picher the heaviest producer of lead and zinc concentrates; the American Lead and Smelting Company; the Mutual Mining Company; and the Bilharz Mining Company. In 1951 the Nellie B. Company was purchased by the American Zinc, Lead, and Smelting Company.[55] The Tri-State Ore Producers' Association in 1946 reported twenty-six major mining and milling companies, fifty-one smaller companies, and twelve mills re-treating tailings.[56]

To illustrate how consolidation continued, Harold Childress, president of the Tri-State Ore Producers' Association, reported to the House Select Committee on Small Business, in conjunction with hearings on the problems of the lead and zinc industry, that in 1953 there were fourteen major lead and zinc producers and sixteen smaller ones in the Tri-State District.[57]

Another interesting development in Tri-State enterprise was the method of marketing lead and zinc concentrates. For all lead and zinc fields except the Tri-State, East St. Louis and New York City were the chief market centers. St. Louis quotations were basing points for domestic production, and New York was a basing point more for foreign imports, and prices there were generally higher because of the freight differential. This ranged over the years from $.035 a pound to almost $.02 a pound.[58] While this was basically the story of national concentrate marketing, much of the Tri-State ore was smelted in Kansas and Oklahoma. The method of actually marketing these ores then differed from that used in other sections of the United States. As late as 1931 the fourteen major smelters maintained ore buyers with offices at Joplin and they generally purchased firm requirements weekly.[59]

The concentrates were purchased directly from the mine in most cases, and the handling and freight charges were paid by the smelter, whereas Western operators paid freight charges. An open market situation existed in the Tri-State District, too. While the St. Louis basing price was used as a starting point, there was competition among the ore buyers who made purchases by bids.[60] Of course the first ore buyers were the land and royalty companies who operated lead smelters on their properties and sold the refined lead to manufacturers. After 1872, with the rise in the demand for zinc, outside buyers came in to purchase blende. Soon, they were purchasing lead ore, too.

This broke the purchase monopoly of the land and royalty companies. The early ore buyer traveled about the mines in a mule-drawn wagon. His practice was to inspect sample handfuls of ore from bins, and by rule of thumb he judged the quality of the concentrates, deducting for impurities and "also a liberal allowance for risks." It was claimed that after the buyer made his purchase and "as soon as he was out of sight, the operator brought out the water hose and turned it into the bin until the ore was saturated. More or less of this water remained to be weighed and paid for as ore."[61]

As the operators formed associations to force higher prices and prevent surpluses, ore buyers gave the ore a more careful scrutiny, chiefly by means of the assay. The general method used by 1900 was to take sample shovelfuls from the loaded ore car. This was mixed together and divided into three parts. One portion was kept aside in case of loss and the other two parts were assayed for mineral content. The results served as a basis for payment.[62]

The practice of weekly sales stemmed from the fact that the district had been a small-operators' haven, and most miners going on a shoestring investment had to have a weekly turn-in of ore in order to survive. In no other mining district in the country were such quick returns possible.[63]

Just as national depressions and foreign imports adversely affected the marketing of Tri-State concentrates, so did the buying habits of manufacturers and labor disorders. In 1903 the United States Steel Corporation, which bought one-half of the nation's smelted zinc, withdrew future purchases indefinitely, since it had sufficient galvanizing zinc for the rest of the year. This forced closing of zinc smelters, the results being felt in the Tri-State District when ore buyers failed to purchase concentrates. This in turn forced the closing of mines and mills.[64] Strikes outside the Tri-State District affected concentrate prices too. The steel strike in 1952 was largely responsible for forcing the price of concentrates so low that operations temporarily ceased.[65]

In addition to promoting domestic markets, the Ore Producers' Association sought foreign outlets. As a result of local publicity and invitations, foreign ore buyers were in the district in the 1890's and early 1900's. In 1890, Vivian and Sons of Wales, England, purchased one thousand tons of zinc blende from the Center Creek Company of Webb City to ship to their English smelters. Vivian paid thirty dollars a ton while local ore buyers were offering only twenty-eight dollars. Some Belgian buyers were also reported in the field.[66] Welsh buyers contracted for fifteen carloads of zinc in 1896. Liverpool, England, buyers took five hundred carloads, and fifteen cars were shipped to Havre, France.[67] In 1901 the Missouri-Kansas Zinc Miners' Association was consigning between five hundred and fifteen hundred tons of zinc weekly to J. Needham Smelters in Antwerp, Belgium.[68]

While corporate financing tied district mining lands into a sort of royalty monopoly, mining up to 1910 was largely in the hands of independent operators. Yet, even independents combined, if in a loose fashion, in order to gain higher ore prices, to obtain favorable legislation, and in general to promote operator welfare. A sagging ore market was a constant menace to the operators, and the chief reason for unified action among the operators was to keep concentrate prices up and develop new markets. Custom or central milling, used briefly in the 1870's, had been abandoned after a short time, and small operators concentrated ore on their own hand jigs or small mills.

Soon, larger operators found it impossible to hold ore for higher prices, since smelter representatives generally could buy sufficient concentrates to fill their needs from small operators who had to sell each week to stay alive. The situation became so critical that in April, 1890, one hundred operators from the camps met at Joplin and formed the Southwest Missouri and Southeast Kansas Lead and Zinc Miners' Association. According to the association bylaws, the association was established "for the purpose of advancing mining interests, and promoting the general welfare of the district by securing concert of action and harmony of feelings on the part of different mining camps."[69]

The association was represented in each camp. Its chief work consisted of regulating the supply of ore through milling agreements when a surplus was apparent.[70] The price-fixing program of the association was quite successful, and, by 1899, landowners and royalty companies were again requiring their small lessees to custom clean their crude ore if they refused to abide by association embargoes on ore sales. Higher ore prices meant higher royalties. Thereby a combination of operators and landowners was able to cut off 75 per cent of the district ore and fix concentrate prices to ore buyers.[71] An added function of the association in 1891 was to free local operators from domestic markets by sending representatives to Europe. They were to seek new markets and advertise the opportunities for investment in the district. Assessments on members paid the cost of maintaining these agents abroad.[72]

By 1900 the combined value of lead and zinc concentrates was running between nine and eleven million dollars annually. Operators worked to maintain at least this level, "a ten million dollar business." This could be accomplished by increasing production and by holding up prices. To accomplish the latter, on one occasion, the association "caused a shutdown of seventy percent of the mills" in the district for three weeks. "This was not a loss of production though," Nicholson added for the benefit of stockholders, "for the mines would have had to shut down anyhow to clean ponds and repair mills, so operators used the shutdown for that purpose."[73]

Several additional instances can be found where the association forced ore buyers to submit to the concentrate price set by operator-members. For example, Jesse Zook, a local statistician, reported that the association in 1905 voted to lower production so as to curtail ore bin supplies, thereby preventing a surplus and sagging prices.[74] A more specialized combination, the Joplin Ore Producers' Association, was set up in 1916. A sum of $750 was subscribed monthly by member producers to employ a secretary and collect useful data for the mine operators. The emphasis of this group was more statistical than price fixing.[75]

The Tri-State Ore Producers' Association was formed in 1924 with headquarters at Miami. A year later the association moved into a new office building at Picher, a more convenient meeting spot. The primary objectives of the association were:

> 1. To collect, compile, and record all available data and statistics touching or concerning the production or consumption of zinc and lead ores. 2. To promote the interests of its members by ethical and lawful means. 3. To cooperate with the authorities of government in the administration of the laws affecting directly or indirectly the zinc and lead mining industry of the Tri-State District, composed of parts of the states of Oklahoma, Kansas, and Missouri, wherever, and whenever such cooperation is desired or requested.[76]

In the beginning the association staff consisted of a secretary and safety engineer, responsible for promoting and coordinating safety programs in the mines.[77] In addition the association held "a weekly roundtable of the metallurgists and mill men of the field . . . at its offices through most of the year. Many and varied problems arising out of legislation, insurance, freight rates, and other matters of economic or technical nature involving the local mining industry as a whole are continually being handled through the Association."[78] Other activities of the association have included health services. In 1927 the producers, in conjunction with the United States Bureau of Mines and the Metropolitan Life Insurance Company, established a cooperative clinic at Picher, Oklahoma, for the purpose of treating silicosis and setting up an educational program to control the disease.[79]

Also, the association from 1924 to 1931 maintained "an industrial welfare department with a director and two nurses with cars who devote themselves to human problems of the working people and families of the mining field."[80]

Several other instances can be found of association and cooperation among operators. Most of the towns of any size had businessmen's clubs. The leading members were operators. The clubs in Joplin and Galena were especially active in promoting markets for district output and gaining greater tariff protection.[81] Mining exchanges, formed more specifically for transactions in lead and zinc properties and royalties, also worked for markets and furnished production statistics. These were situated in Joplin, Galena, and Baxter Springs.[82]

While each of these operator organizations sought markets and better prices for district concentrates, furnished statistics to operators, and, in one case, did some work in employee welfare, they also engaged in lobbying for a number of miner-operator causes, including added tariff protection. Also, when the Quapaw Indian lands were opened to mining, the exchanges lobbied for a relaxation of Interior Department regulations on the use of these lands.[83]

In 1953, when foreign imports of duty-free lead and zinc forced a shutdown of district mines, the Tri-State Ore Producers' Association sent a delegation of mine operators to Washington to appeal for tariffs on foreign imports.[84] Harold Childress, president of the association, appeared before Congressman William Hill's Select Committee on Small Business, where he presented a statement on the district's distress caused by the government's tariff program.[85] Nationally, the district producers were affiliated through the American Zinc Institute, with headquarters in New York City. The organization served as a research, public relations, marketing, statistical, and legislative oversight agency for lead and zinc producers in the United States.[86]

CHAPTER XII

Tri-State District Social Conditions

Tri-State promoters took great pride in announcing the mounting production and profits of the district. They could demonstrate with reports and statistics that annual lead and zinc concentrate production had increased from 25,845 tons valued at $1,019,400 in 1850 to 942,488 tons worth $57,321,173 in 1925.[1] These reports and statistics failed to show, however, the human cost—physical deterioration and sordidness that came to the miners whose labor was in large measure responsible for this splendid production achievement.

Tri-State miners were exposed to every sort of hazard. In excavating a shaft there was always the danger of cave-in. Cribbing, wooden timbers laced together, was used to hold the soil and rock sides in place as the shaft deepened to solid rock. If the cribbing were improperly placed or jarred by blasting, the soil and rock might be freed to cascade down on the workmen. Once the ore vein was struck, drifts, or horizontal tunnels, were driven to follow the deposit. With drifting the hazards mounted. While pillars were used to prop the drift roof at intervals, vast areas of open rock space were exposed. Probably the most common cause of mine accidents were boulders, pieces of rock roof, falling on workmen. And at least one-half of the mine fatalities were from falling roof pieces.[2]

The states of Missouri, Kansas, and Oklahoma each maintained a mine inspector for their respective portions of the Tri-State District. These officials had the function of enforcing their respective state's mine safety regulations.[3] Their primary concern was oversight of underground workings, especially supports and roof conditions. Ventilation was a matter of investigation too. To keep a flow of good air coursing through the shafts and drifts, operators installed ventilation fans,[4] sunk special shafts, cut drifts to openings, and drilled air holes with a churn drill.[5] Roof fall was a perennial problem, however, for an inspector might certify a mine free of hazards, and then stope blasting in another part of the mine later in the day would jar safe roof sections loose to fall on workmen below.

Special workmen called roof trimmers, using long ladders and pry rods, worked to increase mine roof safety.[6] These specialists inspected the roof after each charge was set off. Loose roof was pried down. Then the miners were allowed to enter. Even this precaution did not completely remove the hazard of falling rock.[7] Drift cave-ins were a menace too, and many times miners were trapped underground for days. Joseph Clary was a cave-in victim who was fortunate enough to survive, and his experience produced national attention.

> Sunday, July 30, 1911, a cave-in at the White Oak Mine near Joplin imprisoned Joseph Clary, a miner, at 80 feet depth. . . . Three companions were able to escape. . . . Drilling from the surface over his approximate location began immediately. By Tuesday (fifty hours later), a 5″ hole penetrated the drift, and Clary responded feebly. Telephone communications were set up through the drill hole, and the prisoner was allowed a short conversation with his mother at her home a half-mile away. Pump lines were let down for flooding was a danger, and water was already to his armpits. Food in the form of milk and whiskey was lowered and its effect was soon noted in the strengthened voice of the imprisoned man. He was later provided with elaborate food, a watch, electric light, and newspaper, all of which was lowered through the drill hole. A light rope was left to enable him to signal rescuers on the surface. Work began in thirty minute shifts to open the shaft, and shortly he was reached.[8]

In some of the Oklahoma mines the health of workers was imperiled by underground water strongly charged with hydrogen sulfide gas. Workmen coming in contact with this gas were blinded temporarily if they stayed in the drift beyond a two-hour shift. After surface air circulated through the contaminated drifts, the menace gradually dissipated.[9] Clarence A. Wright noted that most of the district mines were wet, the miners working in drift floors covered by an average of three inches of water. Also, some water dripped from the roof. After a day's work the miners came out with soaked clothing. In 1913 a few mines had special change houses, locally called dog houses, equipped with lockers and showers where the men could change from their work clothes, bathe, and put on dry clothing before leaving the mine. Wright concluded in one of his reports that "health considerations . . . have been neglected to a certain extent by many of the operators."[10] Wright's 1918 study, however, showed that most operators had installed change houses with lockers and washing facilities.[11]

Next to roof falls, the most common cause of mine accidents came from the unwise use of explosives. The only way to excavate the ore underground and free it from its solid rock associate was to drill a pattern of deep holes, stuff these with dynamite, and set off the charge. Early miners frequently used a loading stick with a steel nail for inserting the dynamite.[12] If the nail struck the sides of the powder hole, a spark was produced and often a premature explosion resulted. Also, miners were sometimes killed or maimed by delayed explosions when they had not fired.[13] Wright also blamed high royalties for worker hazards: "Drifts are often left in such condition that the lives of miners are endangered in the rush to increase production and thereby add to investment earnings. High royalties also tend to reduce wages, because lessees are compelled to work the mine as cheaply as possible, and consequently the labor is apt to be less efficient."[14]

In addition to state mine inspectors looking after the safety of workers, by 1925 there were several safety specialists hired by operating companies for this purpose. The Tri-State Ore Producers' Association was the first private agency to promote improved conditions in the mines through its mine safety department. Soon the larger companies employed safety specialists too. Their work was coordinated through the association safety department.[15]

Besides watching the roof more closely, safety engineers saw to it that men using sledges or grinding drill points wore goggles. Lighting darkened drifts with flood lamps also reduced accidents.[16] While undoubtedly there was an element of humanitarianism as well as concern for worker efficiency in its relation to output, one of the admitted reasons for operator emphasis on safety was that, as accidents were lowered, "there would result a lowering in the rates paid for compensation insurance, as these rates were becoming increasingly burdensome."[17]

Above all else the safety programs adopted by the operators stressed dust control. Although water might be up to three inches deep on the drift floor, there was still considerable rock dust in the mine atmosphere. This dust, it was discovered in 1914, accounted for the high incidence of silicosis in Tri-State mines.

Silicosis, or miners' consumption, locally known as miners' con, was not a new disease. Agricola's famous *De re metallica*, published in 1556, and translated by Herbert Hoover in 1912, describes dust-induced respiratory infections as common ailments among central European miners in the fifteenth and sixteenth centuries.[18]

The most serious social problems present in the Tri-State District grew out of this disease. Miners breathing the flint dust produced by drilling and blasting often became infected. Thereby, workmen were predisposed to tuberculosis and other respiratory diseases. In crowded, poorly-housed camps such as Picher, miners' families also became infected with tuberculosis in many cases.

Miners were infected by consumption before 1913, but there is no mention of this condition in the state and federal reports until that time. Wright is the first government writer to note the prevalence of the disease.

> There is usually more or less dust in the air and inhaled by miners, causing lung trouble if breathed for a considerable time, just as lead poisoning is brought about by breathing the fumes at lead smelters. The extent of harm done by inhaling the dust in metal mines is not definitely known, but is believed that much of the lung trouble often known as miners' consumption has been caused by it. . . . The Bureau of Mines, in cooperation with the Bureau of Public Health, is to investigate a number of problems relating to the hygiene of mines and to the mine industry, the investigation being conducted with regard to scientific as well as administrative problems and including a study of the prevalence and prevention of lung diseases among miners.[19]

By 1914 the disease had become so prevalent in the Joplin portion of the district that operators combined to form the Southwestern Missouri Mine Safety and Sanitation Association. Weekly meetings were held to plan a program of education and treatment.[20]

A nurse was hired by the association to visit afflicted miners and give first aid, as well as to instruct members of their families on protective steps. Many operators began to sprinkle the ore face to abate the dust, but the *Engineering and Mining Journal* charged that "the workers were often negligent."[21] Also in 1914, the Jasper County Anti-Tuberculosis Society was organized. It was supported by the operators as well as by general citizen groups. Besides developing a program of education, the society hired an additional visiting nurse and promised another for the Galena field as soon as funds were available. Temple Chapman, a local operator, donated a twenty-acre tract near Carterville for the erection of a "tent city and open air camp for the victims of the disease."[22]

The visit by a unit of the United States Public Health Service, as announced by Wright, in 1913, took place in November, 1914. According to the *Engineering and Mining Journal*, the purpose of the unit, headed by Dr. Anthony J. Lanza, was "to make a survey of hygiene and sanitary conditions in the lead and zinc mines with a view to helping with measures recommended to help miners suffering so greatly from tuberculosis. The statistics are so high on deaths from phthisis during the last year in the Joplin district as to seem impossible. The number of patients treated by the Anti-Tuberculosis Society is now 150."[23]

Before Lanza's time, health activities in the district dealt largely with tuberculosis, which represented a final development in progressive lung deterioration. Most of this started with silicosis.[24] Lanza's study got under way in November, 1914, and continued until July, 1915. Two publications resulted: *A Preliminary Report on Sanitary Conditions in the Zinc-Lead Mines of the Joplin, Missouri, District and Their Relation to Pulmonary Diseases Among the Miners*, published in 1915; two years later there appeared his more definitive work *Siliceous Dust in Relation to Pulmonary Disease Among Miners in the Joplin District, Missouri.*

Lanza's first step was to advertise throughout the district that each miner would be examined free of charge at the Webb City headquarters. In all, 720 miners responded.[25] Each received a chest X-ray and complete physical examination. Lanza concluded that the powdered flint rock formed "a very fine, hard, sharp, and insoluble dust which permeates the underground atmosphere and which . . . is extremely irritating to the lungs when inhaled, causing the condition known as miners' consumption."[26]

Thus for the first time a medical scientist had studied and produced a definite explanation of the cause of miners' con.

Rather than concentrating on treatment, Lanza emphasized the necessity of going to the source of the disease—mine rock dust. To care for those who had also contracted tuberculosis, the Southwest Missouri Safety and Sanitation Association took over the management of a hospital at Webb City.[27]

Lanza's study found that the dust inducing silicosis came from the following underground sources: drilling, blowing out drill cuttings with an air blast, blasting, shoveling, boulder popping, tramming, and pillar trimming. Dust on the mine surface came from dumping ore buckets and crushing mineralized rock in the mill. The amount of dust produced by each source was mechanically measured, and the heaviest offender was found to be the practice of blowing out drill holes with a blast of air.[28]

Lallah Davidson's *South of Joplin*, the *Preliminary Report* of the Tri-State Survey Committee, and Frederick L. Ryan's *Problems of the Oklahoma Labor Market* jumped at the conclusion that dust producing silicosis also coursed through the mining camp atmosphere infecting the wives and children of the miners, too.[29] Lanza's work was available for the consideration of these writers, and, had they consulted it more carefully, they would have seen that only those persons working in close association with the dust, either as miners or millmen, were affected. These writers also would have found that

> mechanically produced, miners' consumption is neither contagious nor infectious. The lung, irritated, inflamed, and injured by the hard-rock dust, becomes fibroid—that is, there is developed throughout it a scarlike tissue, which interferes with the normal elasticity of the lung and prevents proper breathing. The disease may attack one or both lungs, but is usually evenly distributed on both sides. . . . as . . . consumption becomes established there is a gradual impairment of working ability, owing to shortness of breath, with coughing, loss of weight, and weakness. There is a disposition to catch cold easily, and the cold is apt to settle on

> the chest. The intensity of miners' consumption varies, all grades of severity being seen, depending on the nature of the dust and the length of time spent underground, or other extent of exposure. It may persist for years and tends gradually to produce total disability. There is, however, a deeper menace than gradual disability, in that the injured lung, together with the lowered vitality of the whole body, presents an ideal site for the development of tuberculosis. When this occurs, not only is the miner threatened with an early death or at least, a more helpless future; but having now a contagious disease, he becomes a source of danger to others—his fellow workmen and his family. A tuberculosis infection may take place at any stage of miners' consumption.[30]

In order to curb the conditions which produced silicosis, Lanza made the following recommendations for Tri-State operators and miners: (1) improvement of miners' housing so as to remove one of the causes of family infection once the miner father's silicosis had reached the critical stage; (2) wetting down of all mine workings after blasting, using water injection with drilling and no boulder popping on the shift. (3) installation of fans to circulate air within the mines; (4) erection of change houses at each mine, equipped with ventilated lockers and showers; (5) setting up of sanitary drinking fountains and toilets underground. The lack of the latter, he noted, had resulted in a promiscuous ground pollution in the mines; and (6) periodic chest examinations for all workers and their families to check the control rate of both silicosis and tuberculosis.[31] The Missouri Legislature readily responded to his recommendations and passed a mining law in March, 1915, requiring that all lead and zinc producers provide for their workers change houses, sanitary toilets, and drinking fountains. Operators were obliged to drill and blast with sufficient water to abate the dust. Any operator failing to comply was subject to a one-hundred-dollar fine and the closing of his mine by the mine inspector.[32] Kansas and Oklahoma subsequently adopted similar measures.[33]

Wright found health conditions considerably improved in Tri-State mines by 1918.[34] But the number of silicosis and tuberculosis cases remained constant, so in 1923 the Bureau of Mines set out to determine whether the measures recommended through the Lanza study had been followed and to what extent they were adequate in preventing silicosis in the Tri-State District. In 1924, 309 miners were studied for silicosis. Of these, 101 were negative, 114 doubtful, and 94 showed positive signs of silicosis. Consequently, the bureau reiterated Lanza's recommendation of nine years previous, that all miners receive annual physical examinations. To accomplish this the Tri-State Ore Producers' Association, the Bureau of Mines, and the Picher Post of the American Legion set up a clinic at Picher, staffed with a physician and clerk. Within three years more miners were applying for examinations than the limited facilities could handle. Therefore, in 1927 the association, the Bureau of Mines, and the Metropolitan Life Insurance Company established an expanded clinic. The association contributed sixteen thousand dollars annually, while eight thousand dollars each was furnished by Metropolitan and the Bureau of Mines for the clinic budget. A large X-ray room, dispensary, and venereal disease clinic were included.[35]

A total of 7,722 miners were examined in 1926. The subjects ranged in age from sixteen to seventy-eight years, with the average approximating thirty-one years. Of these, 6,495 were between twenty and forty-nine years old.[36] Incidence of silicosis within the early age group was low, probably due to the shorter period of employment, but increased drastically in the subsequent categories.

The 1928 investigation, the first in a series of two, contains some interesting data concerning personal habits of the miners examined.

> The use of tea, coffee, or tobacco apparently has no relation to silicosis or tuberculosis; but after silicosis has developed, the use of alcoholic drink and patent medicines increases decidedly—alcohol from an average of 13.64 to 25% and patent medicine from 3.75 to 12.50%. Other habits, such as the use of tobacco and tea, decline. The men take alcohol and patent medicines to alleviate the symptoms of silicosis so that they can continue work. It is not believed that alcohol and patent medicines are predisposing causes of silicosis or tuberculosis but rather that the men resort, in many instances, to the use of alcohol or patent medicines, particularly the latter, in an endeavor to find a remedy or cure for their disease.[37]

Overcrowding was also a subject of investigation for the Bureau of Mines in its 1928 study. "An inquiry into the sleeping accommodations of men employed in the mines of the Picher district revealed that more than 75% of the men slept in rooms shared by one other person, that about 20% slept in rooms occupied by two other persons, and 10% stated that three or more persons slept in the same room with them. No correlation could be made between silicosis or tuberculosis and these conditions."[38]

The Bureau of Mines published a supplement to the 1928 paper a year later concerning family status and silicosis. The family history investigation involved 8,817 miners.[39] The lowest incidence of silicosis and tuberculosis was among single men, probably due to the age difference and number of years spent working underground. Widowers and divorcés showed the highest percentage respectively.[40]

The 1929 supplement also attempted to show some correlation between silicosis and venereal disease, another serious social problem of the district.

> Divorced men and widowers show the highest incidence of syphilis. This fact may have some bearing on the incidence of silicosis as silicosis has been found to be more pre-

> valent in syphilitics than in nonsyphilitics. The possibility that the wives of these men might have had tuberculosis should be considered, but the personal history given by the men does not reveal this. . . . Gonorrhea was reported by 29.1% of the men and syphilis by 6.03%. Close correlation of the incidence of syphilis with uncomplicated silicosis indicates that syphilis may be a predisposing cause of silicosis. Clinical observation further leads to the belief that silicosis runs a more rapid course in syphilitics. Such cases certainly show a definite tendency to develop lung abcess. To reduce the incidence of syphilis, a venereal-disease clinic was organized in the latter part of 1929.[41]

This study showed that all operating companies were using some form of dust control, including wet drilling and ventilation of underground workings plus sprinkling of ore piles before shoveling. In summary it announced there was "a very definite improvement over previous conditions."[42]

Dust control was improved in 1939 by the development of a device to measure the quantity of rock dust in the mine atmosphere. A norm of five million particles a cubic foot was established. Mine operators sought to hold the count below this. Technicians pointed out that when the dust count exceeded this norm, a silicosis environment was created.[43] Despite these improvements, when Secretary of Labor Frances Perkins held a conference at Joplin in 1940 to determine the status of health and working conditions in the district, it was disclosed that rock-dust control, along with housing, ranked as the leading problems.[44]

By 1929 all employees of those companies cooperating with the Picher Clinic, consisting of 80 per cent of the district miners, were being examined at least once a year. "Those with first stage silicosis are given warning cards, and many leave the mines. Men who continue to work are not recommended for re-employment after they develop second-stage silicosis. This plan offers them an opportunity to leave a dusty occupation at a time when they have the best chance to recover, as they have been informed of their condition and the likelihood of serious consequences if they continue to work under such conditions."[45]

After the 1929 report was published, all applicants for jobs in Tri-State mines were required to undergo a physical examination including X-ray at the Picher Clinic. This agency also made its services available to the families of miners where there was a possibility of contagious disease, especially tuberculosis.[46] Alice Hamilton, a Bureau of Labor specialist on industrial poisoning, visited the district in 1940. Concerning the periodic physical examinations, she found the workers were generally suspicious, and "the men distrust the clinic and mistrust the examining doctors paid by the companies, whose duty it is to weed out the silicotic and the tuberculous, which to the miner spells unemployment and starvation."[47]

The 1929 report recommended that "the cases of silicosis plus tuberculosis appear to do better in the high dry climate of the Southwest." The air of the Tri-State District was described as moist and not favorable for silicosis victims. Not only should the patient move to the dry air climate, but he would "do better with a limited amount of daily exercise."[48]

This recommendation raised new problems for infected miners and their families. Generally lacking the money required for such a move to a healthier environment, silicosis victims remained in the district, eking out a precarious existence by farming on mining land, going on relief, or doing odd jobs such as their physical condition permitted. To add to the family income, the wife and children often went to work. The social problem pyramided, since this meant neglect of home life as well as education for the children. Assuming the miner had savings sufficient to move to the "high, dry climate of the Southwest," the problem of earning a living still confronted him. More often than not his only skill was that of mining, and his physical condition ruled that out. Federal Social Security and state aid to dependent children programs have helped alleviate some of the distress caused by silicosis.

Another industrial disease growing out of the Tri-State metal industry and producing its train of social problems due to worker disablement and consequent family hardship was lead poisoning. Ryan and others have inaccurately attributed this ailment to the handling of crude ore in the mines.[49] Actually there is no basis for this.[50] Lanza's study demonstrated that, although particles of lead were suspended in the air of district mines, "there was nothing at all observed to lead in any way to the conclusion that metallic poisoning was responsible for diseases of miners in the Joplin District."[51]

Lead smelting, one of the district's leading industries, was responsible, however, for producing this disease. A concern for treatment of lead poisoning came as early as 1877, thirty-six years earlier than for silicosis. Some of the early remedies included washing smeltermen's hands in petroleum three times daily. This removed the lead dust and "the fatty substance in the oil prevents absorption of the lead" into the system.[52] Another preventive, developed by a Belgian physician, was known as Melsen's Treatment. It consisted of giving doses of iodide of potassium to smeltermen. This chemical purged the system of poisonous lead fume.[53]

A simple remedy was discovered in France in 1878. It consisted of giving infected smeltermen large quantities of fresh milk daily. This reduced the pain of lead poisoning, and, if the patient were treated soon enough, he was restored to full health. French smelter operators gave their workers extra money to buy a quart of milk a day, and it was found effective in preventing the disease in many cases too.[54]

Those smeltermen, painters, and others who become victims of lead poisoning, sometimes called lead colic or painters' colic, still use milk as a deterrent to poisoning.[55] According to modern medical research, lead is absorbed through the lungs, skin, and digestive organs, and, while this absorption takes place readily, throwing off the lead is rather difficult.

> One form of lead poisoning begins suddenly. In such cases a large amount of lead is absorbed rapidly, usually through the stomach. Pain in the abdomen, vomiting, and collapse are symptoms of this acute form. Severe colicky pains and rapidly developing anemia with a typical white waxy color of the skin is frequent. Examination of the blood in such cases shows a peculiar appearance of the red cells. A blue line around the gums is also an important symptom in many cases of lead poisoning, but it is not always present. The most important symptoms of the slow or chronic poisoning are paralysis, usually of the arms, the brain. Headache is common and patients are frequently either depressed or excited.[56]

The menace of lead poisoning in the Tri-State District smelters was diminished by an effective program of precautionary measures. Eagle Picher, operator of the Joplin and Galena smelters, required all employees to receive periodic physical examinations. Smeltermen had facilities to clean themselves before mealtime. For sure prevention, even the workers' fingernails were cleaned before handling food. The Joplin and Galena smelters also had shower facilities for workmen after completing a shift. And company rules prescribed that each worker wear a respirator mask before entering the plant.

Tri-State counties took at least limited steps to improve the health of miners and their families through county health units.[57] Jasper County had a sanitarium at Webb City to care for tubercular patients, and the county unit had an annual budget of $23,000 to improve local health conditions. In 1939 the Tri-State counties, through their respective state governments, entered into a health compact looking to the improvement of conditions in the mining field.[58]

In Ottawa County, however, the Tri-State Ore Producers'

Association did most of the work on community health and sanitation until 1949, when the county health unit largely took over. The association established an industrial welfare department in 1931. This consisted of a director and two nurses who visited miners' homes and assisted in the improvement of community health.[59] The association's Picher Clinic, concerned with examining miners for employment, ceased operations in 1932. The task of examining mining employees and applicants for employment was continued until 1939 by the Tri-State Ore Producers' Association under the name of Tri-State Industrial Board. Later, miners were examined in local hospitals.[60] The association continued to look after the health of Picher residents, however, through the employment of a visiting nurse, Ruth Hulsman. Her work, extending over a period of twenty-three years, reflected the problem of health in the Picher field. Describing her activities, a writer for the *Daily Oklahoman* said:

> For 23 years, the needy people living among the lead and zinc mines—the people with disease, with destitution, with problems too big for them—have been coming to Ruth Hulsman, welfare nurse. Or she has been going to them until it has totaled more than 75,000 visits. . . . "When I came here in 1926 . . . epidemic raged. The schools would be closed with outbreaks of typhoid or diphtheria or smallpox. Now we seldom have a case of typhoid or diphtheria so that if we do, I feel personally disgraced. It wasn't always easy. You don't go around sticking people with a hypodermic needle to prevent a deadly epidemic. Often you battle ignorance, superstition, indifference, and fear until it's years before you awaken people to the crying need that an ounce of prevention is worth a pound of cure." . . . There's always prevalent tuberculosis, constantly a public health problem in the mining section. In 1943, at the peak of war prosperity, of 1,473 persons, mostly school children, given tuberculin tests, 295 or 20% were positive. She can't begin to tell how many she has taken each year to the Talihina sanitarium.

> . . . You cannot begin to recount the things Ruth Hulsman has done beyond the call of duty to help people. She's seen that school children had shoes, books, and glasses. She's verified their need for free lunches. She's helped them get ice through the summer, garden seed, a cow, fruit jars, milk for babies. If the kind of help they need isn't in her budget she knows where to find it. . . . She has probably looked upon more helplessness, more destitution, more sickness, than any other person in this district. Sordidness and wretchedness, human misery in its lowest dregs, have often been her workshop.[61]

Despite the attempts of local county health units and the Tri-State Ore Producers' Association to control silicosis and tuberculosis, this disease combination persisted as the district's primary social problem. In 1941, Dr. Joseph Spearing, director of the Cherokee County Health Department, said that for the preceding year, "Cherokee County reported more active cases of tuberculosis than any other county."[62] In 1951, Cherokee County's death rate from tuberculosis was six times greater than for the rest of the state of Kansas.[63]

For years, Ottawa County, Oklahoma, had the unenviable recognition of having the highest tuberculosis mortality rate in the nation. While in 1942 Ottawa County's record was 201.5 deaths for one hundred thousand people, this was reduced to 86.9 in 1951.[64]

Undoubtedly one of the chief reasons for the decline in the tuberculosis mortality rate in the Tri-State area was due to the application of improved health and control measures.[65] Another explanation is found in the reduced number of workmen employed in the mines. The Tri-State mine labor force declined from 11,117 in 1924 to 4,100 in 1953.[66] This can be accounted for through increased mechanization in underground workings requiring fewer miners for a shift and to declining production throughout the Tri-State District.[67] As this region's deposits became increasingly marginal after 1930, the scope of operations decreased proportionately.

Fewer men working underground meant less silicosis and therefore a lower tendency toward tuberculosis. Yet the high rate of silicosis in the past seeded the region with tuberculosis and a bumper crop of infection was reaped, not only among the miners, but their families and nonmining residents as well. In 1936 the Kansas State Board of Health conducted a tuberculin testing program among school children in the Cherokee County towns of Baxter Springs, Treece, Columbus, Galena, and Riverton. A total of 991 students and 89 teachers participated. An alarming 36.34 per cent positive reaction was registered by all age groups, and among those over twenty years of age the per cent of positive reaction was 53.9.[68] Then, in 1943, a similar testing program was conducted among Picher, Oklahoma, school children. Tuberculin tests were given to 1,473 persons, and of these, 295 or 20 per cent were positive.[69] This portended a bleak health future for many young people of Kansas and Oklahoma portions of the district.

CHAPTER XIII

Tri-State Labor Relations: The Early Period

The Tri-State District was a famous mining center for over a century. It was the scene of many changes in method, capitalization, and personnel. Of these, however, the most fundamental was in personnel, since the human element was more directly involved. It required nearly a century for the Tri-State miner to evolve from a versatile, independent operator to a specialized worker seeking a daily wage and affiliated with a union. Traditions die slowly though, and while a definite labor-management pattern had emerged by 1953, there were those miners who persisted in carrying on the "poor man's camp" idea, seeking their security not in labor unions and collective bargaining but rather in their own labor and skill in independent operations. Yet each year this became more the exception than the rule.

In the period before 1910, when the small operator prevailed and mine labor crews were small, there was little division of labor. Each man who worked in the diggings was versatile. This early miner was a small capitalist since his grubstake investment made the undertaking possible. He was an amateur geologist, knowing something of rock formations, ore-bearing strata, and the relation of surface configuration to ore bodies. He was acquainted with the elements of prospecting, knowing how to construct a churn drill, operate it, and sample the cuttings. He could run air drills, load the holes with dynamite, and set off the charges. Also, he operated pumps to keep his mine free of water. This early miner knew how to crib a shaft and construct props underground to make the mine roof safer. He shoveled the ore, served his turn as hoisterman, and milled the crude into concentrates. He repaired his equipment when required and served as carpenter to build mill and hoist housing.

As mining became more mechanized and was conducted on a larger scale, labor crews increased in size. There developed a greater division of labor in the interests of efficiency. Workmen became specialized. Each specialist performed a single phase of mine operation, and his function was identified by a particular name. Often this designation took on a local flavor. Some of the more picturesque ones include: (1) hoisterman —the man who operated hoisting machinery for lowering and raising the orebuckets or cans. Sometimes these were called tubs. (2) roustabout—an unskilled laborer who carried equipment and performed common labor tasks in the mine and on the surface. (3) screen ape—workman stationed at the ore screen with a heavy hammer to sledge the ore-bearing rock into pieces small enough to pass into the skip below. (4) roof trimmer—safety specialist who inspected the mine roof after each blast to determine its safety and pried loose rock down by working from a long ladder when required. (5) pig tail—one who trammed the ore cans from the drift to a lay-by or switch point where a string of cars were collected and pulled to the shaft for hoisting. (6) bruno man—underground worker given the task of pulling fine ore piled on the stope from blasting. (7) mule skinner—operator of electric locomotive or driver of mules used to tram ore cans from the lay-by to the shaft for hoisting. (8) tub hooker—workman stationed at the juncture of the drift and shaft to grasp the hoist hook, affix it to the loaded ore can, and give the lift signal to the hoisterman. (9) cull man—surface laborer who sorted the contents of the ore can, discarding the barren rock and sending the ore-bearing material to the mill. (10) mucker—underground workman who hand shoveled the ore into cans. (11) cokey—another local designation for shoveler. (12) cokey-herder—

supervisor of a crew of shovelers, sometimes called a shoveler boss. (13) powder monkey—miner skilled in the use of explosives in underground excavation.

Besides these specialists, each mine had a crew of electricians, blacksmiths, machinists, pumpmen and pipe fitters, night watchmen, carpenters, trackmen, mill men, churn-drill operators, metallurgists, and management staff personnel, including the safety engineer and mine superintendent.[1]

The Tri-State mine labor force from 1848 to 1954 was almost exclusively male. Women and children were used in the 1870's to wash dirt from the ore and sort it by hand,[2] but there is no record of employing either women or children for underground work. Dolph Shaner, a district pioneer, recounts that a man and wife mining partnership operated a claim on East Joplin Hill in the 1870's. They had made a rich strike, and in later years their discovery was known as the Dutchman's Mine. "The husband did the digging, while his wife, in a Mother Hubbard apron, and sun bonnet, did the windlassing."[3]

It is impossible to state, or even estimate, the number of men employed in mining in the district at any one time. Even in recent times a few miners shifted from the wage earner to small operator class and back again into the wage earner group if their grubstake was exhausted before ore in paying quantities was found. This mobility was not nearly as widespread as in the pre-1910 era,[4] when practically every wage earner worked only long enough to save sufficient funds to begin prospecting again. James Bruce, a local mining engineer, noted this independence and enterprise among Tri-State miners. "This is a poor man's camp and almost every miner who has spent any number of years in it . . . has . . . at some time owned a prospect, and has made at least some attempt to organize a company to secure a lease and try his luck at finding diggings."[5]

There was a constant movement back to farming, especially in the Missouri section of the field. This adds to the difficulty of estimating the number of Tri-State males earning a living through mining. Several studies on district labor through the years[6] show that the peak year was 1924, when 11,187 miners were employed. This dipped to 1,800 in 1934. An explanation for the small number of men employed in 1934 in relation to the number shown for 1924 is partially explained by the national depression. After 1935, when about 3,500 were employed, increased mechanization took place, further reducing the number of workers required to man the district mines. Added to this, between 1941 and 1945, was the effect of the selective service system on district males. In 1953, 4,100 miners were at work.

The number of hours worked in the mines varied too. When miners were working in partnership operations, it was not unusual for them to labor from dawn until dark in their feverish efforts to score a strike. If a miner were working for wages before 1900,[7] he customarily put in nine hours a day, six days a week. Smeltermen put in twelve hours a day, six days a week in winter and eight hours in summer.[8] In 1899 the Missouri Legislature passed an eight-hour law for miners.[9] Like other miner benefits produced by political action, the Missouri Legislature and the eight-hour law were condemned by the operators through their spokesman, the *Engineering and Mining Journal.* The editor declared in 1905: "Much trouble will be produced in the Joplin District. The customary working time has been nine hours. . . . Joplin has presented the best example on record of the advantages to labor of labor increasing its own efficiency, and it is a pity that this has to be disturbed by short sighted legislation."[10]

The Oklahoma Constitution of 1907 prescribed eight hours a day for underground work and forbade the employment of males under sixteen in the mines. Females of all ages were forbidden to engage in underground work.[11] Kansas also restricted the number of hours in the mines to eight in 1917.[12]

The Lead and Zinc Code of the National Industrial Recovery Act had the effect of making the eight-hour day uniform in the mines,[13] and the Fair Labor Standards Act of 1938 made the forty-hour week, with time and a half for overtime, mandatory in the district.[14]

Just as hours varied in the district mines, so did wages. Regardless of the wage levels, however, miners were always paid weekly in cash and not in company chits redeemable at a company store as was the case in some other mining regions.[15] Although the Tri-State wage level generally increased through the years, it was below the national average daily rate paid to lead and zinc miners. In 1929 the local wage of $4.00 a day compared to $4.90 national average.[16]

The Tri-State average wage scale trebled between 1920 and 1950, but the rate differential among the labor types remained about the same. All workmen in the mines were paid on an hourly and daily basis except cokeys or shovelers. Their work was paid for on a piecework basis, presumably in order to stimulate greater production, since the output of a mine was reckoned in the number of cans of crude ore produced in a day's time. Their pay for each can ranged from ten to twenty cents.[17] A good shoveler could make more money than any other worker in the mines. Otto Ruhl, a local mining engineer, noted that while this piecework system afforded the greatest production, it had its drawbacks. He wrote in 1911: "While the shoveler today can, if he will work a full week, exceed the salary of the superintendent of the mine, the system has almost outgrown its usefulness. It keeps men from becoming proficient machine drill men. Why become skilled when shoveling yields by far the best wages? Thus those indisposed to shoveling gravitate to machining and thus cripple the efficiency and effectiveness and production of the mine."[18] With the modernization of mines, the hand shoveler was largely displaced by mechanical loaders.[19]

One of the most perplexing questions arising out of Tri-

State labor history is the relative lack of an aggressive union movement there until 1935. Immediately on the northern fringe of the district in Crawford County, Kansas, the coal miners carried on vigorous union activity with considerable violence in the 1920's. Yet the Tri-State District was generally quiet. Oklahoma's coal fields were the scene of labor activity and violence, too. Yet Ottawa County was quiet until 1935. Many reasons have been advanced to explain this socioeconomic phenomenon. Some writers point with some validity to the mixed mining and farming carried on in certain portions of the district. Others claim with less validity that the lack of any appreciable foreign-born population explains the absence of labor trouble. A few attribute this anomaly to the miners' intelligence, and there are observers who explain it in terms of the "poor man's camp" tradition. The matter is too complex to be explained satisfactorily by any single factor.

Undoubtedly, one cause of the relative lack of labor action in the district was that most of the labor force was recruited locally from the small farms of the Ozark hill country.[20] Wiley Britton, an early settler who witnessed the evolution of the district from a hand windlass to electrical hoist field, noted that, for the most part, local men were employed in the mines.[21] Charles W. Burgess, a local mining engineer, reported that "labor generally is plentiful, especially in the winter months, as many of the men engaged in farming come in and work in the mines."[22] Agriculture, then, furnished some security for the worker if the mines shut down.

One of the most popular reasons advanced to explain the lack of labor union activity was the absence of a sizeable foreign population in the district. Since a national stereotype of the early 1900's was to associate labor strife with the number of foreign born in the community, many writers on Tri-State affairs marveled at the local labor peace and accounted for this condition by the absence of foreigners. Burgess was among the first writers to claim that the district's labor peace was due to the fact that "no foreigners are employed."[23] Walter Williams' study of the district in 1904 praised Joplin as "a city of self-made men, nearly every one of the moneyed citizens having made his fortune there. They are largely American born and American educated. For thirty years there has never been a strike or labor disturbance to mar the native's good name."[24] Lane Carter, a Chicago mining engineer, found that "the camp is singularly free from labor troubles, and although agitators come along periodically and try to stir up trouble, miners' unions which have caused such trouble in many other camps, have not damaged the district yet. The visitor is surprised to find that there are practically no Negroes at work in the mines of Joplin. What Negroes are there are employed in other occupations than mining. There are scarcely any foreigners."[25] A. Hecksher, a local investment promoter, concluded that "few labor problems" were present in the Tri-State District because the leasing system gave "every man a chance . . . and the miners are almost all exclusively native, the foreign element being less than two percent of the population."[26]

Clarence A. Wright's *Preliminary Report* of 1913 noted that district miners were nearly all "Americans . . . and usually efficient,"[27] and the *Daily Oklahoman*'s feature article on the district in 1922 observed that "the miners are all American and free from the influences of labor organizations."[28] Most local operators, as well as their spokesmen—for the most part mining publication journalists—indoctrinated workers with the pride of being American and the evils of the foreign element. Yet there were times when these same operators promoted foreign immigration. This represented an odd contradiction. The foreign element was disparaged because of its alleged tendency to produce labor strife. On the other hand, rising labor costs or shortages of American miners caused several operators to promote importation of foreign workmen. In 1909, Jesse Zook, a local statistician, announced that operators faced with rising costs must "increase production to keep profits up. This is in no way aimed to force a lower wage scale, but simply gain a larger output. The supply of American shovelers is about all utilized. Some operators favor hiring Italian laborers for this work."[29]

A year later, when the Federated Mining Company tried to import Italian laborers, "the foreign labor was not allowed to remain by the American workers."[30] Otto Ruhl, a leading district investment promoter, warned in 1911 that while many companies were seeking to import foreign labor, such action was inadmissible since "American labor is so much more efficient that any change would entail a revolution in mining and milling experience. Besides, American laborers are jealous of the exceptional conditions here and they resent any influx of foreign element. So far, it is impossible to get foreigners into the field."[31] Commenting on the war labor shortage in the district during 1918, Ruhl noted that "shovelers are scarce. The draft is taking them for they are the younger men. Older men do the work and are less efficient. This cuts down on production. Some mines tried to hire Mexicans, but white men refuse to work and threaten trouble for the Mexicans."[32]

J. B. Rhyne, writing on the district in 1929, observed that it was "unique in that it is almost entirely made up of native born whites. There are no Negroes [in Picher] and very few foreign born whites."[33] The census reports, too, show the Tri-State District had less than 2 per cent foreign born in the population.[34] Crawford County, Kansas, adjacent to Cherokee County on the north, and immediately on the fringe of the district, had a much higher percentage of foreign born, 16 per cent in 1910 and 7 per cent in 1950.[35] Until 1921 this coal-mining region boasted the most aggressive labor activity in Kansas. It is strange that the neighboring lead and zinc district should "have been little affected by this labor conflict."[36] It was less because of the foreign element in Crawford County, however, and more because of the nature of coal mining

—larger operations where no "poor man's camp" tradition could develop. Crawford County lacked the resources to sustain a farmer-miner population like those which Jasper and Newton counties possessed. Also, the people were natives in the lead and zinc region and knew better how to subsist, while the Italian and Slav coal miners in Crawford County lacked the know how.

Therefore, Crawford County miners were less independent and more easily influenced by union organizers. Also, general conditions were worse in the company-controlled coal-mining towns of Crawford County than in the lead and zinc communities like Joplin, Webb City, Neosho, and Granby. Tri-State operators looked askance at the coal-producing center, and, when a strike there resulted in a slight migration of coal miners to the lead and zinc district in search of employment, it produced alarm among the operators since the influx was sometimes a "disturbing element."[37] Even down to fairly recent times, Tri-State vanity has manifested itself on the foreign worker issue. One of the best instances can be found in a statement by the Rev. Cliff Titus before Secretary of Labor Frances Perkins, who investigated Tri-State labor and housing conditions in 1940. Titus at the time was pastor of the First Community Church of Joplin, probably containing the wealthiest and most exclusive clientele in the district. The reverend was speaking out against the interference of Washington in local affairs, and serving, perhaps inadvertently, as apologist for the health and housing problems of the district.

> We have here a group of laboring men in our Tri-State District which is all American. Now we say that with no aspersion at all on foreigners, of course, but it happens to be that the men who work in the Tri-State area are Americans, white Americans. They have quite a tradition behind them; they are pretty independent; they don't like to be bossed much. They have done pretty much as they pleased; if they didn't like the ground boss in one mine they would go across

> the road and work for a ground boss in another mine. When conditions are good they receive good pay.[38]

A much more valid explanation of labor apathy in the district is found, not in the lack of a foreign element, but in the wide range of opportunities available to miners until recent times.

Early miners were not only workmen, but operators as well, since the scope of operations were small. Because of its wide opportunities, one commentator explained: "The Joplin district can't be organized because . . . a majority of workers always are carrying on an interest in some prospect and as soon as it is developed, miners leave their jobs and go to work in their own diggings, thus becoming employers of labor, under which conditions they have no use for the union."[39] Writers constantly referred to the fact that "local miners are as a rule intelligent and seemingly contented. Strikes are unknown."[40] Carter surmised that "the fact that so many working men are leasing ground on their own account is one of the reasons for the satisfactory labor conditions in the district. It is in a measure, the profit sharing idea."[41] Irene Stone, daughter of an early mining promoter, added that, while "the landowner may not reap so great a profit," from the small operations system, "there are no strikes or other disagreeable features arising from employing help."[42]

While no labor organizations were successful in the district until 1935, several instances can be found of at least incipient miner organization looking to the improvement of earnings and working conditions. The earliest union of miners took place at Joplin on March 7, 1872. A mass meeting occurred at this time, and the following resolutions were adopted. According to notice, the citizens and miners of the Joplin lead mines assembled in mass meeting February 17, 1872. John Riley was elected president and John W. Howe, secretary.

> Rules: *Article 1.* No miner shall hold more than one lot nor company of miners more than one lot to every able bodied

man. *Article 2.* Each and every miner shall hold his claim and he shall work one able bodied hand for each lot so held. *Article 3.* Whenever a miner shall fail to comply with the above articles, any man, after twenty days of failure may declare the lot forfeited, and work it himself, except when sickness or other providential accident shall prevent the worker. *Article 4.* All disputes arising between miners shall be settled by a jury of six miners, to be chosen by the parties in the controversy who shall be under oath, to decide the case according to the evidence and these rules. *Article 5.* The decision of the jury shall be final on the case, unless appealed within ten days, to a higher court. It is hereby recommended that the parties in the controversy sign a bond to abide the decision of the "miners court" before going into trial, then the verdict shall be final. *Article 6.* It is hereby made the duty of the miners to elect one of their number a miners magistrate, to swear the witnesses and jury, and perform all the necessary business appertaining to the legal trial of cases and causes arising under the miners' code, and who shall keep a record of all proceedings, and who shall hold his office for a term of one year. *Article 7.* Whenever any of the miners shall become dissatisfied with any of these rules or resolutions, they shall publish a call for a mass meeting for six days previous, and submit to the decision of such meeting. *Article 8.* The miners magistrate shall be paid two dollars for each case tried, and each juror shall receive one dollar for each verdict rendered. Said cost to be paid by the party in the wrong.[43]

The foregoing miners' code was drawn up largely by independent operators of small means, and, while for the most part they found the "poor man's camp" atmosphere congenial, they had complaints.[44] These are reflected in a set of resolutions amended to the miners' code.

Resolution 1. That it is the inherent right of the miner to prescribe the terms and conditions of mining on ground once thrown open to mining work. And we hereby wish it distinctly understood that we deem ourselves grossly

> wronged by certain persons claiming the right to buy our mineral at their own prices. *Resolution 2.* That we hereby declare it to be the duty of the miner to sell his mineral to the man who will pay the highest price for it. Always reserving the rent for the owner of the land. *Resolution 3.* That we deem ten percent of the price of mineral sufficient rent.[45]

At Granby in March, 1874, the Miners' Benevolent Association was formed. This organization was established to provide relief for "disabled or distressed miners." A fund was set up and was administered by a committee of miners. Hardship cases were reported to the committee, investigated, and if the situation justified aid, the committee authorized disbursement of funds to the applicant. Money for the fund was furnished by assessments against members. All lands at Granby were controlled by the Granby Mining and Smelting Company and leased to miners, who were required to sell their mineral to the company. The Granby Company withheld ten cents for every one thousand pounds of lead mined and two cents on each two thousand pounds of zinc for the fund and agreed to match one-half of the amount paid in by the miners.[46]

Other evidences of miner organizations and unified action can be found. There is at least one instance of miners striking, rioting, and destroying property before 1900. On the evening of July 20, 1874, fifty masked men used a quantity of dynamite to destroy the Hannibal Lead Works, an adjunct of the Picher Lead Company at Joplin.[47] This climaxed a three-day demonstration of miners who "crowded the streets and held numerous meetings" of protest against royalty payments and mineral sales. Land and royalty companies were collecting a 20 per cent royalty from their miner lessees and requiring the workmen to sell their mineral to the companies rather than in the open market.[48]

The miners contended that a 10 per cent royalty was sufficient and that they would receive more than the traditional twenty-five dollars for each thousand pounds of mineral if they could sell in the open market. Since the Picher Company was the most reluctant to grant the miner demands, its properties suffered most heavily from the demonstrations.[49] These acts of violence, though, did not win a permanent redress of grievances for the miners. Royalties were reduced briefly from time to time, but the trend until 1910 was for a steady increase until, in some cases, the tribute was as high as 50 per cent. As to the sale of ore concentrates in the open market, the land and royalty companies abandoned the practice of requiring lessees to sell to them after ore buyers came into the field in 1875. It should be noted that the miners engaging in these demonstrations were not, for the most part, wage earners, but rather operators. Little hired labor was used.[50] Most of the early miners were operators, working on a partnership basis.

The most frequent combination was two, allowing one worker underground to dig and one on top of the ground to hoist and clean the ore. Instances can be found at Joplin of as many as nine men working on shares in the same mine.[51] At Granby there is a record of several twelve-member partnerships working on the shares and at least one case of thirty-two sharing miners.[52] A few mine operators became very wealthy through most fortuitous strikes,[53] and some early-day miners made as much as fifty dollars a day from their diggings. Most operators, however, barely made the equivalent of wages.[54] The Missouri labor commissioner found that miners customarily worked for wages just long enough to grubstake themselves, then they secured a lease and began prospecting, thus becoming operators. If no minerals in paying quantities were found before their savings were exhausted, the operators took jobs in the mines for wages again. Some miners moved in and out of the operator class twenty times within their life span.[55] The fact that a Tri-State miner could move with relative ease from the wage earner to capitalist status accounts in great measure for the lack of a vigorous labor movement in the district in the period before 1910. By this date, deeper mining in the Kansas and Oklahoma portions of the field, as well as the Webb City-Duenweg-Oronogo district in Missouri, requiring heavier capitalization and more modern equipment, severely restricted the scope of small operations. In recent times, however, shallow scattered deposits around Spurgeon and Racine, Missouri, defied large modern mining enterprise, and small operators could still function as was the case a century ago.

CHAPTER XIV

Tri-State Labor Relations: The Recent Period

In spite of the glowing reports concerning Tri-State labor relations,[1] there was an undercurrent of discontent. On several occasions, Tri-State miners combined and conducted mass demonstrations to protest conditions without the benefit of labor organizations.[2] The fact that operators sometimes met their demands evidently convinced many miners that there was little need for union affiliation when one could gain redress through local demonstrations.

The *Engineering and Mining Journal* explained that the reason Tri-State miners were not active in the labor movement was because, while strikes at Joplin mines might occur, these disturbances were "due entirely to local causes and never due to or undertaken by a union organization."[3] And in commenting on the district's peaceful labor record, the editor added that while "Joplin workers will strike for betterment . . . they seldom will listen in great numbers and act on the directions of agitators. . . . Joplin is noted for its freedom from Unionism and independence of the men, who have been notorious strike breakers in Colorado."[4]

Nonunion demonstrations in 1905 and 1906 typify the tendency of Tri-State miners to seek redress of grievances. Workers left their jobs in 1905 to protest the failure of operators to observe the Missouri eight-hour law for mines.[5] The operators promised to respect the law, and work was resumed. The *Engineering and Mining Journal* evidently saw an omen in the 1905 demonstrations and reported that "during the last week in November there was incipient labor trouble. This was

Webb City Mining Camp, Missouri. (*Courtesy Joplin Mineral Museum*).

Picher Mining Camp, Oklahoma. (*Courtesy Joplin Mineral Museum*).

Chitwood Mining Camp, Missouri. (*Courtesy Joplin Mineral Museum*).

Open-pit mining, Oronogo Mining Camp, Missouri. (*Courtesy Joplin Mineral Museum*).

Leadville Hollow Camp, Missouri. (*Courtesy Joplin Mineral Museum*).

Duenweg Mining Camp, Missouri. (*Courtesy Dorothea Hoover*).

Opening a mine in Missouri. (*Courtesy Joplin Mineral Museum*).

Shoveler crew in the Lucky Strike Mine, Duenweg Mining Camp, Missouri. (*Courtesy Dorothea Hoover*).

Chitwood Mining Camp, Missouri. (*Courtesy Joplin Mineral Museum*).

Oronogo Mining Camp, Missou

ourtesy Joplin Mineral Museum).

Stotts City Mining Camp, Missouri. (*Courtesy Joplin Mineral Museum*).

Prosperity Mining Camp, Missouri. (*Courtesy Joplin Mineral Museum*).

Rich Hill Mine, Missou

ırtesy Joplin Mineral Museum).

Swindle Hill Mine, Missouri. (*Courtesy Joplin Mineral Museum*).

something previously unheard of here, but it didn't amount to much."[6]

Early in 1906 the workers were out again, this time seeking nine hours pay for eight hours labor. After a respectable period of refusal, the operators granted this too. This made unsuccessful "the unionization of the district," which had been attempted by the Western Federation. But the *Engineering and Mining Journal* warned that "efforts toward that are still being made, and the poison has been introduced which manifests itself in discontent and decreased efficiency."[7]

A strong reason for district mine owners meeting the non-union demands of 1905 and 1906 was that the Industrial Workers of the World had established a local in Joplin in 1906. In the year of its founding, the IWW local held a joint convention with the Jasper County Socialist Party, another agency that had a militant labor policy.[8]

Actually, however, the IWW was not the first labor organization in the district. Approximately two hundred local miners were associated with the Knights of Labor in 1889.[9] There is no evidence that this affiliation produced a positive labor program for district workmen. Then in 1893, three hundred district miners affiliated with the Convention of Miners with headquarters in Kansas City, Missouri. This union called a strike in August of 1893 as a demonstration of sympathy to bolster the cause of the Kansas coal strike in Crawford County. The *Engineering and Mining Journal* reported that "at a few of the lesser mines, some of the men went out, but at the great mining centers, the workmen refused to strike."[10]

Frederick Ryan's study of the district labor movement reveals unions "existed only through sufferance and as long as they were quiescent."[11] Thus, by mine owners meeting worker demands, the Western Federation of the IWW lost out in the district in 1906. But it returned in 1910 and organized a new local: "The Joplin local is one of about 200 branches of this organization in the country and its members have the interchangeable card system and have many benefits which individuals do not have. The local is affiliated with the Joplin Trades Assembly and meets at Labor Headquarters once a week."[12]

This union's six hundred district members went out on strike in June, 1910, in protest to working conditions and for a one dollar a day raise. Surprisingly, no less spokesman than the traditionally operator-slanted *Engineering and Mining Journal* came out in favor of the strikers:

> There is a moderate strike in the Joplin District. Irrespective of the merits of the present case, there is no mining district of the United States where we are so well pleased to see the miners get all they can. It is not generally known that the mining conditions of the Joplin District are highly dangerous and unsanitary, but such is the fact. The mines of the district are operated to a large extent upon a relatively small scale and with a good deal of carelessness. The ore is extracted in large chambers untimbered, and falls of roof are a common source of accident. The annual fatality rate of four per 1,000 is high. . . . We should not be surprised if mining in the Joplin District were really as dangerous as work in an arsenic factory.[13]

Commenting on these district labor developments in 1910, Otto Ruhl, a district mining engineer and investment promoter, wrote: "The district has always been recognized as non-union and an unorganized camp. In recent years, the miners have shown a growing tendency towards some cooperation among themselves and this year saw a respectable number of miners organized into a union. Its further growth means increased wages and higher living costs. Also, an attempt to import foreign labor would result in solidifying this union sentiment, a thing producers are fearful of."[14]

Operators were able to foil the 1910 protest with strikebreakers. Local grocers cooperated by refusing to extend further credit to strikers. But the Western Federation shortly recovered and, boasting three thousand district members in 1914, again called a strike for improvement of conditions and pay raises of up to one dollar a day. The crews of only two mines went out immediately, and the movement appeared to have been beaten.[15] Then in June, 1915, all three thousand members struck for the union demands set forth during the preceding year. Like a great labor army, the miners marched through the district urging nonunion members to join the strike. Mine owners, according to the *Engineering and Mining Journal*, countered by shutting "down for two weeks, and say they will stay closed indefinitely rather than deal with the Western Federation which is behind the strike."[16]

Within three weeks the operators had broken the strike. Union leaders harangued the miners to stay out, promising that the union could raise $100,000 to extend the strike. But the operators pledged that workmen would be paid a sliding scale of wages based upon the average price of ore, thereby gaining the raise demanded.[17]

The ill-fated strike of 1915 completely demoralized the cause of Tri-State mine unionization for twenty years. In 1916 the *Missouri Trade Unionist* reported twenty-six unions functioning in the district, but not a single one represented the miners.[18] Several unsuccessful attempts were made after 1916 to organize unions. An instance was the attendance of three thousand miners at a meeting held under the auspices of the International Union of Mine, Mill, and Smelter Workers at Picher, Oklahoma, in April, 1925. No successful organization came from this meeting. This unionization attempt, according to Ryan, was negated by the operators who "made organization unattractive by raising wages as soon as union representatives entered the field."[19]

At least one attempt was made to organize Tri-State smeltermen into a union before 1935. In 1911 the Federal Labor Union organized a local among smeltermen employed by the Eagle Picher Company. But after eighty or ninety workmen joined the local, the *Missouri Trades Unionist* reported that "it was found that spies were reporting on the meetings and the local decided to cut them out for a time. But the local is expected to be revived in the near future, bigger and stronger than ever in its history."[20]

Labor activity in the district had a radical fringe too. From 1905 to 1907 the Socialist party had an organization in Jasper County. The party sponsored a weekly newspaper from October, 1906, to April, 1907, the Carl Junction *Socialist News*, published in the mining town of Carl Junction, six miles northwest of Joplin. The Jasper County Socialist party sought to organize a mine labor movement, and its editorials encouraged miners to join the local unit of the Industrial Workers of the World. Its editor announced on October 2, 1906, that "the Progressive Local No. 240 of the Industrial Workers of the World meets every Thursday night at the Joplin Court House at 7:30. All visiting brethren are invited to attend."[21] Both miners and their wives were eligible for membership in the party. Monthly dues were fifteen cents for men and ten cents for women.[22]

The county chapter of the party sponsored coworkers, called "comrades" by the *Socialist*, who traveled throughout the district, speaking in the mining camps on behalf of the Socialist movement. The paper carried testimonials on why people became Socialists.[23] Samples of these include "James Adair of Zincite is in the field teaching Socialism. His salary is $50.00 per month,"[24] "Reverend Adair will discuss Socialism from the Biblical standpoint at the Joplin Local of Industrial Workers of the World meeting in October," and "Comrade Phil Callery spoke in Stotts City September 25th to a large audience. The meeting was orderly. Dr. Hunter, a Congregational minister and five other pastors in Carthage are studying Socialism."[25] The *Socialist* filled its columns with invective toward Tri-State mine operators. An especially virulent sample, published in January, 1907, was titled "Jasper County

Wage Slaves." "Joplin is the place where the capitalists throw the 'hooks' into the factory hand and take from him the largest share of what he produces. It is the place where gay and festive capitalists who hate unions, union wages, and conditions, reside."[26]

An interesting phase of Tri-State labor relations was the recruitment of local workers by Western mine operators for strike-breaking purposes. Ruhl admitted that "Western mining companies use the Joplin district as a recruiting field when they inaugurate a non-union labor system."[27] And the *Engineering and Mining Journal* advised its readers in 1915 that "Joplin is noted for its freedom from unionism and independence of the men, and they have been notorious as strike breakers in Colorado and other Western states."[28] As late as 1939 the Joplin *Globe* and Joplin *News Herald* carried under their "Help Wanted" columns requests for skilled miners to migrate to Colorado, Idaho, and Utah. These ads were subscribed by Western mine operators.[29]

In spite of the hostility of Tri-State operators, a vigorous labor movement was finally under way in the district in 1935. Stimulated by the collective bargaining clause of the National Industrial Recovery Act,[30] the American Federation of Labor launched a union drive in the Tri-State District in late 1933. Within a year its affiliate, the International Union of Mine, Mill, and Smelter Workers, had established locals at Joplin, Galena, Baxter Springs, and Picher. Tom Brown, the union's agent, recruited four thousand members from the district's miner population.[31]

Brown found the region economically distressed. A large number of workers were unemployed and on relief in proportion to total population. The high number unemployed and on relief reflected the one-industry character of the district. Ottawa County, with a total population six times less than Oklahoma County, had an unemployed and on relief count equal to Oklahoma County. Greene County's population exceeded that of Jasper County. Yet the latter surpassed Greene in the number of persons on relief and unemployed.[32]

A mine labor force of over ten thousand in 1925 had shrunk to three hundred in 1934.[33] Ryan's labor study shows that conditions improved somewhat, for the four thousand lead and zinc miners in 1935 since 1,800 had full-time employment. He added that "the remainder pick up odd jobs around the area, work on W.P.A., live on relief, or secure odd days of work in the mines. . . . In October, 1934, one of the largest mines paid the workers on the average of $8.35 a week, the workers putting in less than half time. Of nine mines, four of them month after month showed employment of about two-thirds time, with average wages of $15.00 to $16.00 per week. The zinc miner, working full time, can earn about $21.00 per week."[34]

In March, 1935, the International Union of Mine, Mill, and Smelter Workers undertook to gain recognition from district operators. Also, a raise in wages was sought since 1935 saw the reopening of most of the local mines with an improvement of the concentrate markets.[35] Four overtures by the union for collective bargaining were rejected by the operators. The fourth refusal, in May, 1935, resulted in a strike called by the union for May 8, 1935.[36] Mines and smelters were picketed by AFL union members, and on the morning of May 8, 1935, mining and refining ceased in the Tri-State District.[37] No violence was in evidence in the early period of the strike.[38]

The operators through the Tri-State Ore Producers' Association persisted in their refusal to recognize the union and declared the policy of the district operators would continue to be open shop.[39] On May 21, 1935, a back-to-work movement was launched. An estimated fifteen hundred AFL members and nonunion men, feeling the hardship of the strike, staged a mass parade in Miami.[40] This appeared to be a spontaneous movement, lacking direction and purpose except that of returning to work. Actually it was a well-planned movement with skillful direction from management officials, as the National Labor Relations Board pointed out in subsequent hearings. It appears that, since the workers wanted a union, local mine management would give them one. Instead of a labor organization dominated by the workers and their agents through the AFL, though, it would be a company union operated by the mine owners for their mutual benefit.[41]

The architect of Tri-State labor relations for the next four years was F. W. "Mike" Evans, a mine operator. This "friend of labor" was a member of the Tri-State Ore Producers' Association and operated the Craig Mine, employing thirty-five men. Between 1932 and 1935, Evans produced $385,000 worth of ore which he sold to the Eagle Picher Company. Evans owned the Connell Hotel in Picher, had a partnership in a firm selling secondhand cable, owned a business building in Picher, as well as a night club and pool room, and was part owner of the Picher Ford Agency.[42] In addition to all these managerial connections, "this remarkable man," according to Malcolm Ross, a National Labor Relations Board investigator, "also owned a whiskey still in an abandoned mine . . . said by revenue officers to be the largest ever raided in those parts, a distinction for which Mike served a stretch in jail."[43]

The next step in breaking the AFL strike following the first back-to-work parade took place on May 25, 1935, when twenty-eight men assembled near Quapaw, Oklahoma, to form a counter union to the AFL. This group, consisting of mine superintendents, ground bosses, and other supervisory personnel, was led by Mike Evans. These planners decided to stage additional back-to-work demonstrations and build sentiment for the projected union. A rally was held on the same day at the Miami Fairgrounds, attended by approximately 250 miners. Evans presented a back-to-work petition addressed to local operators for the miners' signatures. This document had been drafted by Glenn Hickman, ground boss for the Black Eagle Mining Company and later a public relations executive for the Eagle Picher Company.[44] Within two days, 3,125 signatures had been placed on the petition.[45]

On May 26, one thousand miners assembled at the Miami Fairgrounds for the purpose of taking further action on the back-to-work plan and to hear an address by Mike Evans.[46] In his speech, Evans suggested the possibility of forming a union "to aid in opening the plants and mines and to end the strike." Another mass meeting was planned for the following day, May 27, to elect officers. This session, attended by an estimated three thousand miners, resulted in the formation of the Tri-State Mine Metal Workers Union.[47] A common name for the new organization was Blue Card Union, from the color of membership booklets. Mike Evans was elected president and Glenn Hickman was named secretary-treasurer.

The same meeting saw the adoption of the Blue Card Union Constitution and By-laws. This document made no mention of collective bargaining and pledged its members not to strike. Only operators or mine executives could serve as union officers and as members of the twelve-man executive committee. The constitution, published in the National Labor Relations Board *Hearings* in 1939, showed the aims and objectives of the new organization. These were in part:

> To unite the metal, mine and smelter workers; to promote their general welfare and advance their interests, social, moral, and intellectual; to protect their families by the exercise of justice, very needful in a calling so hazardous as ours.
>
> Persuaded that it is for the interests both of our members and their employers that good understanding should at all times exist between the two, it will be the constant endeavor of this organization to establish mutual confidence and create and maintain harmonious relations.
>
> Such are the aims and purposes of the Tri-State Metal, Mine and Smelters Union.[48]

Following the formation of the Tri-State Union, the strike's first labor violence occurred. This instance set off a series of disorders that were to terrorize the district for the next two years. A delegation of AFL members called on Tri-State Union President Evans at his headquarters in Picher to protest the strike-breaking tactics used by the new organization. When Evans appealed to the Ottawa County sheriff for protection, the law officer was assaulted by AFL members. Thereupon, the 3,150 Tri-State Union members were rallied by Joe Nolan, another local operator and vice-president of the new group, into a "pick-handle brigade." Their weapons were short lengths of seasoned hickory, tapered like a baseball bat, and about two pounds in weight, normally used as handle replacements in miners' picks.[49]

Nolan's brigade was so successful in avenging the attack on Evans' headquarters that the Blue Card Union came to use the mass demonstration and attack to crush all future AFL resistance. Nolan's pick-handle brigade broke up AFL meetings all over the district. Locals at Joplin, Galena, Baxter Springs, and Picher were raided on numerous occasions. When AFL men took steps to protect themselves, they were arrested by local law enforcement officers for rioting.[50]

Units of Kansas and Oklahoma National Guardsmen were stationed in the district during 1935 to insure that striking AFL men would not interfere with the opening of mines and smelters.[51] Operators signed agreements with the Tri-State Union, in effect themselves, and ignored completely the AFL's attempt to gain recognition. Also, the operators, especially those at the Eagle Picher Company, the biggest enterprise in the district, gave generously to the Blue Card Union. Up to July 5, 1935, Eagle Picher alone contributed fifteen thousand dollars to the Tri-State Union treasury.[52] Much of this money, according to the National Labor Relations Board testimony, went to pay the cost of squad cars and armed patrols.

Evans and the Tri-State Union took over the task of local

> law enforcement. The record abundantly shows that Evans, with the acquiescence and aid of Sheriff Ely Dry . . . distributed deputy commissions to members of the Tri-State Union. Purportedly to protect property, an elaborate organization was set up whereby "squad-cars" were sent throughout the area to patrol the highways and the mines. These activities were directed by Mike Evans . . . with headquarters at the Connell Hotel in Picher, Oklahoma, a hotel owned by Evans. These squad-cars contained four or five passengers, all paid by the Tri-State Union and equipped with guns and other arms. Between 75 and 100 men were on the squad-car pay roll. The record abounds in incidents of lawless violence committed by the squad-car men. These squad-cars cruised about the neighborhood, stopping all cars suspected of containing International members, watching houses of International members during the night. . . . An International member harassed by the squad-cars described, "There wasn't no sleep them days." Further, the evidence shows that the squad-car men beat up members of the International, and brought them to Evans' headquarters at the Connell Hotel where Evans . . . and local law-enforcement officers sat. . . . The early organizational activities of the Tri-State Union were confined almost wholly to the activities of the squad-cars, to pickhandle parades, and to violence and threats of violence in order to pave the way toward a mass resumption of work. This campaign was financed in very substantial part by the respondents [Eagle Picher Company].[53]

The National Labor Relations Board findings charged that a "reign of terror" was spearheaded by local thugs and criminals. Mike Evans' personal bodyguards consisted of Sylvester Waters, identified by the Oklahoma State Bureau of Criminal Investigation as "Missouri Criminal Number One," and Ray Jamison, who boasted a long criminal record in Oklahoma.[54] Tri-State Union meetings were dominated by the executive committee consisting of twelve operators and management executives. Joe Nolan, leader of the pick-handle brigade and known as the pick-handle king served as a sort of brutal sergeant-at-arms during union meetings.

A sample of how union business was railroaded through was presented in the National Labor Relations Board testimony. "After Norman read the . . . Tri-State Union Agreement, Nolan said, 'Every man . . . that agrees to this agreement, arise to your feet.' . . . They stood up by the thousands. Then after a whispered conference with Norman, Nolan announced, 'Anybody that is contrary to this agreement stand up and get knocked down, please.' " No one stood up. The meeting lasted for four hours, only forty-five minutes or an hour were devoted to union affairs since the speeches were "very short." The remainder of the meeting was devoted to orchestra music.[55]

The Tri-State Union's time and energy were not used to gain better working conditions, higher wages, and collective bargaining for members. Rather, the rank and file were constantly indoctrinated to be hostile toward the AFL, they were hauled about in trucks to break up AFL meetings, and they staged demonstrations through their pick-handle brigade.[56]

Board investigators stated that previous to parades or attacks on AFL meetings, the Blue Card Union Executive Committee saw that Tri-State members had plenty of food and whisky. They added that there was no doubt that the executive committee, consisting of mine executives and operators, ran the union. The Tri-State Union Constitution provided that "the exclusive management of the union shall be vested in the executive committee of twelve members, which committee shall pass upon all matters pertaining to the affairs of the Union, and its decision by a majority vote of the committee shall be final and binding."[57]

If some rank and file member happened to raise a voice of protest to a ruling or decision made by the committee during union meeting, Kelsey Norman, the union counsel, "told the protestors to sit down and shut up."[58] The executive committee spent most of its time conferring with district operators on anti-AFL strategy and passing on the qualifications of applicants for union membership. While operators claimed that the district was to remain "open shop," the facts contradict this.[59] Before a non-Tri-State Union member could go to work, he first had to go to the mine foreman where he was seeking employment and apply for a rustling card.[60] He then presented the rustling card to the union executive committee. If the applicant had no rustling card, he was not eligible to appear. AFL members could apply, and they stood a good chance of election to membership if their union books showed no picket stamps.[61]

This was designed to keep the "undesirables weeded out."

> The Tri-State Union devised an elaborate system of passing upon applicants. The first step was to require immediate resignation from the International and the turning over to the Tri-State Union the applicant's International Membership Card. If the applicant refused this, he was precluded from further consideration. Thereafter, the applicant's name was presented to the membership for consideration. "Ballots for protest" were supplied all members. The next step was to parade all applicants across a platform before the membership so that the members could "have a look" at the applicants before they voted. Prospective members were asked whether they had picketed, how long they had been in the International, and whether they would defend the Tri-State Union with pick-handles. Three protests excluded the applicant. Even if he was approved at this stage, the executive committee could thereafter reject him. If rejected, the applicant might renew his application in 90 days if during that period he stayed away from the International Hall.[62]

Economic necessity drove many AFL members into the ranks of the Tri-State Union.

> Conditions with labor, the workmen that were out of work, were becoming desperate. There is not much else in this community on which the working man can live except the mines, and that is practically the support of other industry and business in the district. And the men had been on reasonably low wages and, in fact, very low wages up to the time of the strike on account of very low market prices for concentrates. And being down, being out of work a week or two weeks, brought most of them to destitution.[63]

Others joined because of the confusion in national labor organizational affairs during 1935. John L. Lewis, head of the International Union of Mine, Mill, and Smelter Workers affiliate, had sought to accomplish a wider extension of industrial unionism through the parent American Federation of Labor organization. But the AFL voted at its national convention in 1935 to follow a policy favoring the craft unions. Rebuffed, Lewis shortly led a secession of eight international unions from the AFL and formed the Committee for Industrial Organization. Thereafter the International Union of Mine, Mill, and Smelter Workers was affiliated with the CIO. In the Tri-State District considerable confusion grew out of the changeover from AFL to CIO. Former AFL members found it difficult to understand the shift in allegiance, and the company-dominated Blue Card Union undoubtedly benefited from this confusion.

Communities and families were further split by rival affiliation, either Blue Card or International CIO. This animus extended even into the play group. Lallah Davidson, a district visitor in this period of labor strife, reports the following taunt hurled by Blue Card Union children at their CIO playmates:

Heigh Ho, Heigh ho,
I joined the C. I. O.
I give my dues to the Gawddamned Jews
Heigh ho, Heigh ho.[64]

And adult poetry was about as complimentary:

The head of this Union, his name is Mike,
He and Joe Nolan were in the lead of the Parade
with pick handles when we broke the strike.
Mike said let the yellow bellys stay on relief,
And drink their dried milk and eat canned beef.
Because our little Union is just doing fine.
Let the rest of the strikers stay on the soup line.
They are losing their cars and selling their hogs.
For the International Union has gone to the Dogs.[65]

The Tri-State Union succeeded completely in its primary purposes, that of opening Tri-State mines and breaking the International Union's strike. But labor trouble was looming on the national horizon for the operators and their strike-breaking tactics. In 1935 the Wagner National Labor Relations Act became effective.[66] Kelsey Norman, a Tri-State Union attorney, consistently advised his clients that the act was unconstitutional and would be so declared shortly, and the district operators need not be bound by the requirements of collective bargaining and other labor rights provided in the law.[67] The United States Supreme Court, unimpressed by Norman's gratuitous ruling on the Wagner Law's validity, upheld the measure as constitutional on April 12, 1937.[68]

On the same day, Kelsey Norman and Mike Evans signed an agreement with G. Ed Warren, an American Federation of Labor Representative from Tulsa, affiliating the Tri-State Union with the American Federation of Labor. This represented an odd move since, as long as the American Federation of Labor contained the International Union of Mine, Mill, and Smelter Workers, it was the target of Tri-State Union abuse. National Labor Relations Board testimony shows that the Tri-State Union declared in 1935 that the AFL was "infested with reds and Communists," and "William Green and his dirty, blood-sucking leeches . . . have the interest of only one small minority at heart, namely the racketeers who run the American Federation of Labor."[69]

But two years later, when the Wagner Law was validated, the Tri-State Union, aware of the penalties of a company union, sought the respectability that would come from a national affiliation. Yet Tri-State Union Executive Committee members admitted at the labor board hearings in 1939 that they privately affixed conditions to the new Blue Card Union tie and that they told their members:

> Technically we are now operating under the general policy laid down by the American Federation of Labor; but really we are still the Blue Card Union in every respect and as such we intend to remain. If the recently negotiated affiliation means the sacrifice of a single iota of confidence we have established with the employers of this district, it is better that we repent our action and fight to the end as an independent union.[70]

On November 8, 1937, the International Union of Mine, Mill, and Smelter Workers, CIO, filed a complaint with the National Labor Relations Board charging that the Eagle Picher Company and other district operators were engaging in unfair labor practices forbidden by the Wagner Act.[71] An earlier complaint filed on May 23, 1936, failed when Eagle Picher brought a bill of injunction against the National Labor Relations Board. The injunction was not dissolved until May 27, 1937.[72]

The charges of unfair labor practices included one that the operators formed and dominated a labor union and used it to "exercise violence and armed force against persons and the property of persons who were or are members of the International Union." Also, the companies were accused of making membership in their Tri-State Union a condition of employment and discriminating in the hiring of International members.[73] The operators denied these charges. Hearings before the regional examiner, with headquarters at Joplin, began on December 6, 1937, and lasted until April 29, 1938. The operators persisted in denying the charges, but evidence subpoenaed by the board, including Tri-State Union records, pointed heavily to the contrary. An interesting facet of the hearings occurred when Glenn Hickman, secretary-treasurer of the Blue Card Union, was instructed to deliver the records to the Joplin hearing. The proceedings show that Hickman was reluctant, but finally brought some records to the building where the trial examiner sat.

> Hickman . . . [kept] them in his car, which he had locked, and [offered] to produce each specific item as called for. He was then directed to bring in the entire mass. Hickman returned to the hearing with the announcement that the car had been stolen. Testimony of a detective who investigated this incident disclosed that Hickman's car had been burned, destroying all the records. The detective's testimony established that the car had been fired from inside, and that the outside had been damaged only by the heat.[74]

Nevertheless, sufficient records and testimony were procured to enable the examiner to proceed. He found that the operators had denied collective bargaining rights, discriminated against workers for membership in the International Union, and were largely responsible for the recent labor violence. Back pay was ordered for 206 Eagle Picher employees so treated.[75] An additional 193 Eagle Picher workers were to be reinstated by the company. The examiner's report was submitted to the National Labor Relations Board in Washington on August 31, 1938. This agency handed down a decision October 27, 1939, sustaining the examiner's report. The Eagle Picher Company exercised its right of appeal to the United States Circuit Court of Appeals as guaranteed by the Wagner law,[76] and a decision was rendered on May 21, 1941. The court in its review noted that "the only function of the court is to determine whether the board, acting within the compass of its power, has held a proper hearing, made findings based upon substantial evidence, and ordered an appropriate remedy."[77] The respondent, Eagle Picher, had due process at all times, ruled the court, and it took judicial notice of the violence and coercion used by the operators and their Tri-State Union. In sustaining the decision of the National Labor Relations Board, the court added that there "can be no doubt . . . the Tri-State Union . . . was dominated and supported by Eagle Picher."[78]

The Blue Card Union, robbed of company support after December, 1939, faded rapidly out of the labor picture. Only on one other occasion was it used to counter the CIO. In 1940 the Tri-State Zinc Company discharged ten miners for membership in and attendance at CIO meetings and agreed to rehire them only if they resigned from the CIO and joined the nearly defunct Blue Card Union. Subsequent National Labor Relations Board testimony disclosed that the company also warned all employees that CIO affiliation would mean an automatic discharge, adding that they could better insure their future employment by joining the Blue Card Union. The CIO brought charges of unfair labor practices against the Tri-State Zinc Company before the National Labor Relations Board in 1941. Hearings got under way in early 1942, and a decision was handed down in March of the same year. The board found the Tri-State Zinc Company guilty of unfair labor practice as charged. The company was ordered to cease promoting the Blue Card Union and to restore the ten employees to their former positions with back pay.[79]

This decision ended the life of the Blue Card Union in local labor relations. The International Union of Mine, Mill, and Smelter Workers, CIO, came to dominate the Tri-State labor picture. Thereafter, operators refrained from interfering in union affairs, but they regularly voiced resentment against the CIO and denied that it represented the views of district workmen. One instance occurred in October, 1939, when Secretary of Labor Frances Perkins conducted an investigation at Joplin concerning working and housing conditions in the Tri-State District.

One of the witnesses appearing before the secretary of labor was James Reed, the district agent for the International Union of Mine, Mill, and Smelter Workers, CIO. Reed testified that district working conditions and housing for miners were submarginal. He added that, as "a representative of the workers in the area," he was competent to appeal for improvement of conditions through a compact with operators. He closed with the remark that "it is with a great deal of pleasure and good feeling that I am here this afternoon, and able to speak with the Secretary closer than at the end of a pick-handle."[80]

Evan Just, secretary of the Tri-State Ore Producers' Association, followed as a witness, and he flatly denied that Reed and the CIO represented the "real workmen." "This is not entirely a pleasant thing to say, but I want it to go into this record that I am speaking for the operators of this district, and I don't propose to enter in behalf of these operators into some broad compact with this union that only represents an extremely small minority of people in this district's workers. The real workmen in this district are not represented here today, to speak of, and I think that should go into the record."[81]

The Bureau of Labor Statistics made a series of studies on the extent of unionization in nonferrous mining, which includes lead and zinc, between 1940 and 1944. As of August, 1941, the bureau found that

> unionization was found to be relatively extensive in the Western States, whereas in the Tri-State District, few workers were found in unions. Information obtained on extent of unionization indicates that in the non-ferrous-metal mining industry as a whole somewhat less than half of the workers belong to unions. The majority of these are members of the International Union of Mine, Mill and Smelter Workers, a C.I.O. affiliate, which is the principal union in the metal-mining field.[82]

This report disclosed that while nearly 50 per cent of Western miners were union affiliated, only 7 per cent of the Tri-State miners were union members.[83] The bureau's next report on the extent of unionization published in June, 1943, showed that while 60 per cent of the miners of the Western states were members of unions, only 23 per cent of the Tri-State miners were unionized.[84]

Yet this represented an increase of nearly 20 per cent over 1941. A subsequent investigation conducted in August, 1943, showed that 25 per cent of the Tri-State miners were "employed in mines and mills operating under union agreements."[85] In the district after 1940, union organizers were able to gain contracts with specific producers rather than on a districtwide basis. Nonunion mine operators generally met the union scale, thereby negating at least one cause for their workers joining a union. The Nellie B. Mining Company, one of the district's larger mining enterprises, employing 350 miners, exemplified this tendency to pay the union wage, yet it did not "operate under a union contract," according to the Joplin *Globe*.[86]

A brief strike at Eagle Picher in 1948 illustrates another method used by operators to break strikes and demoralize Tri-State labor unionization attempts. The chief issue of the strike was seniority rights. The company claimed the exclusive right to determine qualifications of men working in company installations. E. C. Mabon, company personnel director, explained that while Eagle Picher "did not desire to affect the seniority rights or job security of any man," the company must operate efficiently.[87]

Poor union discipline helped break this strike too. A back-to-work movement among CIO members, the same device that emasculated the 1935 strike, was responsible for bringing the demonstration to an end, according to the Joplin *Globe*.

> Unless an agreement is reached in negotiations . . . reports have been current in the mining field . . . that a "back-to-work" movement may get under way. It is becoming common and general knowledge throughout the mining district that a large majority of the workers thrown out of work for a month due to the strike are anxious to get back on their jobs, and that only a small minority group, mostly former employees at the Central Mill, or International Union leaders, have blocked efforts toward settlement of the strike because of differences in the so called "fringe" issues involved.[88]

Any strike is effective only as long as the rank and file membership supports the collective bargaining policies of its union officials. While Tri-State workers showed a willingness to follow their labor leaders in seeking improved conditions and protesting abuses by striking, the period of resistance was never long. Poor union discipline and a remarkable lack of stamina posed the greatest problems for professional union men in the Tri-State District. Yet, in spite of these problems, organized labor persisted in its efforts to get a vigorous, disciplined union movement under way there.

Tri-State workers were represented by the International Union of Mine, Mill, and Smelter Workers, CIO, until June, 1949. In that year the National Labor Relations Board sponsored an election to determine the future bargaining agent for district union members. This was held between the CIO and the United Cement, Lime, and Gypsum Workers, American Federation of Labor. The latter won the election[89] and represented Tri-State lead and zinc miners until the spring of 1953.

This period was "unmarred by work stoppages." Eagle Picher, with seven hundred employees, signed a contract with this union. Evidently, however, the members felt the AFL affiliate too peaceful, so a new election for a bargaining agent was held in April, 1953. Eagle Picher workers chose overwhelmingly the United Steel Workers of America, CIO, as their bargaining agent. Cal Davis, president of the outgoing AFL union, when questioned about his defeat said, "the workers wanted a change."[90]

The change referred to was quick in coming. On June 20, scarcely two months after the CIO returned to the district, a strike was called at all Eagle Picher operations in an effort to gain a forty-eight-cents-an-hour increase. The company had offered four cents an hour.[91] Other union demands included paid vacations and seniority benefits. The Miami *Daily Record* noted that "many employees are planning to travel elsewhere to seek employment for they do not expect a quick settlement."[92]

This strike of seven hundred Eagle Picher workers spread throughout the district and closed virtually all other mines. The explanation for this is because the strike included Eagle Picher's Central Mill at Commerce, Oklahoma, and most non-Eagle Picher operators brought their crude ore to the Commerce plant for concentrating. The Eagle Picher Company raised its original offer of four cents to fifteen cents an hour on October 27. Local CIO officials mailed ballots to members seeking a majority opinion on the company's gesture. This was rejected.[93] Finally, on December 21, 1953, the union voted to end the strike and accept Eagle Picher's October offer of a fifteen-cents-an-hour increase.[94]

Central milling in the late thirties was hailed by metallurgists and mining engineers as a very great step forward in enhancing Tri-State mining efficiency and production. An inadvertent by-product of this innovation was a similar improvement in the cause of organized labor. Centralization of milling technique, as demonstrated by the Eagle Picher Central Mill at Commerce, Oklahoma, ramified into the labor movement. Most operators transported their crude ore to Eagle Picher for milling. The ore was valueless until milled. Thus, when the Central Mill was closed, so was the district. The mill, operated by union personnel, was closed by the 1953 strike. Virtually all district mines were closed. Union as well as non-union men were unemployed.

In 1953, 25 per cent of the district's four thousand miners were union members. The number of unionized miners increased since centralization of operations produced a much greater interdependence. In former years, if some mines and mills closed because of labor activity, others remained in operation, and strike effects were not felt district-wide. But in recent times, by unionizing a key installation like the Central Mill, the effects of a strike were felt throughout the district.

In spite of this wholesome portent for the future of district organized labor, however, that old menace to the cause of miners' unions, poor member discipline and its corollary, the back-to-work movement, loomed in the 1953 strike. The following extract from an "open letter by a striker," published in the Miami *Daily Record*, pointed this up.

> School will start in about five weeks. Our children need their teeth put in shape. Money will be needed for school books and clothing. Also, for payments on our cars so that we will not lose what we have invested in them. Also, money to pay our rent with, for television sets, radios, iceboxes. In fact, we must have an opportunity to earn. Places where we inquire about work turn us down. They say, "You are a striker. I can't use you." Some men are too old to find jobs elsewhere . . . . So we, as employees request negotiations get underway at once so that an early agreement may be consummated. We request that the parties concerned attempt to end their differences.[95]

CHAPTER XV

Tri-State Society

Tri-State mining society was unique. It existed in a socio-economic region superimposed upon the boundaries of Missouri, Kansas, and Oklahoma. These boundaries were superficial. Almost the only times that state jurisdiction entered the picture was for purposes of taxation, elections, and mine inspection. The people in the district circulated through portions of the three states in the quest for mineral wealth. It was common for people to live in one state, to work in another, and to shop for consumer goods in still a third. They thought of themselves less as Missourians, Kansans, or Oklahomans, and more as the citizens of an economic region which had geological rather than political boundaries. The limits of the regional ore bodies were the only boundaries they knew.

Tri-State mining residents were, for the most part, unlike other Missourians, Kansans, and Oklahomans, and there was little state pride and much regional pride. In order to account for the fact that Tri-State residents were somewhat unlike people in other portions of their respective states, one must consider the peculiar conditions present in this region: The local geology, especially the shallow ore deposits; the perennial haste exhibited by miners, sometimes called the Joplin Colic; the geographical background of most of the people, chiefly Ozarkian; the risk involved in mining investment, producing a sort of gambling instinct; and the personal hazards inherent in mining. These factors combined to produce a kaleidoscopic social pattern and a host of social problems.

Literature concerned with social conditions and problems

in the Tri-State District is limited. Extant materials on this subject are often fragmentary and highly inaccurate. It is amazing that private as well as governmental facilities produced such an abundance of research and published literature on improving mining practices and yet neglected for so long the district's human element and its improvement. This is especially so when one notes that a necessary part of local industrial well-being is the welfare of those who dig the ore and mill it for market.

The first official interest in social conditions in the Tri-State District was shown by Dr. Anthony J. Lanza in his studies of health and sanitation in 1914.[1] The U.S. Bureau of Mines made two subsequent investigations, one in 1923, another in 1928.[2] These studies supply the most fruitful source on social conditions.

Private interest in local social conditions came in the 1930's. The first and most unusual piece of journalism on the district was Lallah Davidson's *South of Joplin*, published in 1939. The author claimed to be a product of the Tri-State mining camps. By her own admission, she taught school at Central City, one of the famous but short-lived Missouri mining communities near the Kansas line. After several years in the East, Miss Davidson returned to the district at the height of labor strife in 1935. Any connection between what actually went on and Miss Davidson's account is purely coincidental. Apparently seeking to make a case for the miners and their industry-derived social problems, she succeeded in producing a ridiculous comedy of "life in the Tri-State diggings," through a kind of amateurish first-person muckraking. The Tri-State Survey Committee, a group of New York social researchers, produced its *Preliminary Report* in 1939. This was more careful and objective than Davidson's attempt, yet its analysis left much to be desired. The *Preliminary Report* lacked knowledge of local detail and, made by outsiders, was a sort of superficial investigation.

Malcolm Ross, an investigator of the National Labor Relations Board, was in the district during the period of labor strife. Several chapters of his *Death of a Yale Man* are devoted to a description of those socio-economic problems he found in the Tri-State District. Ross alone showed devotion to facts and avoided exaggeration. His work is as professional as the other two are amateurish.

Despite research limitations, however, one can glean here and there a glimpse of what life must have been in the local mining camps. Until after 1900, miners worked a nine-hour day, six days a week and had Sunday off. While many farmed, gardened, and cared for livestock in their free time, others hunted and fished in the streams and timberland near the mines. Some patronized the town saloons, gambling dens, and houses of ill repute. Joplin, Galena, and other camps had abundant facilities to satisfy those who sought fast entertainment.[3] At Granby, gambling and prostitution became such a problem that the town council in 1873 forbade anyone to run games of chance or a "bawdy house" within the city limits on penalty of a fifty-dollar fine.[4]

While the consumption of whisky in the camps must have been considerable, most of the miners were reported to be temperate. The Missouri Labor Commissioner noted in 1887 that miners who drank to excess found it difficult to get a job, and in the hiring of workmen, married men received preference over single men.[5]

Miners showed a considerable interest in baseball and boxing. A large portion of the space in early issues of the Joplin *Globe*, Joplin *News Herald*, Granby *Miner*, and Picher *King Jack* was devoted to coverage of athletic events. The report of an unusual Granby baseball game was chronicled as "The Granby Daisy Cutters played the Neosho Club and were defeated by a score of 27 to 64. The material of the Daisy Cutters is superb, and relying on sheer nerve, they neglected their practice."[6] Another diversion among miners was to "group together, exchange experiences, and stories, and this if not watched, is done on company time, and at great cost to the labor efficiency."[7]

One of the favorite miner pastimes was to celebrate a new strike. When a prospector made a discovery of ore, the men of the camp turned out to share the glad tidings. Women and children vacated the streets as they were "pretty rough sometimes." The celebration started with a single-file parade of the miners, winding snakelike in and out of stores and saloons and back onto the street, the participants "singing, whooping, holding their picks . . . and shovels high, ringing cowbells and dragging tin cans." Some carried burning torches. Along the course of the parade, the miners "bought candy, cigars . . . and plenty of whiskey."[8]

The highlight of the week's activities for the miners and their families was Saturday night. According to Walter Williams:

> Saturday night in Joplin is a sight worth going miles to see. All the banks of the city are kept open from 7 until 8, and over $100,000 is paid out in several counting rooms. Then the operators receive pay for the week's turnin, and miners and other laborers are paid their week's wages. From 8 o'clock until midnight, the stores are crowded with people making purchases, paying the week's grocery bill, laying in supplies for next week, and swapping experiences. Fully one-fourth of the week's business in the stores is transacted on Saturday night.[9]

Lane Carter, a mining engineer from Chicago, was another observer impressed by local social behavior of the miners. "On Saturday nights or Sundays," he reported, "if one walks through the crowded streets of Joplin and mingles among the miners, one will hear little foreign talk. Plain 'United States' interspersed with a few emphatic 'cuss words' of Cornish origin is the language of the men."[10]

Law and order for the control of mining camp society posed

a problem. After a few months of turbulence, however, the "respectable people" were able to gain an upper hand, and lawlessness largely disappeared. Joplin had its "reign of terror," until Murphysburg and Joplin were united.[11] In each camp, municipal government was soon established, its functions carried on by public officers, and the citizens were represented through a city council. Picher, Oklahoma, was an exception. Modern Picher has municipal government, but for the first three years of its existence, it had, according to the *Daily Oklahoman*, a "feudal organization":

> The company (Eagle Picher) employs a deputy sheriff who has authority to enforce regulations where needed. The social organization is rather feudal in character. The whole town of Picher . . . is built on land leased by the company. As the company's representative, Mr. Bendelari is sort of an overlord, a court from whose judgment there is no appeal. He administers the law of the land. Community differences which inevitably arise are brought to him for adjudication when the litigants are unable to effect a settlement themselves. The company control of the land vests its representative with the power to make his judgments binding. Anyone who refuses to accept the court findings can be dispossessed of his home. Rarely is this extreme penalty imposed. Chief offense against which there is no compromise is infraction of the bone dry law. Eviction is promptly decreed against the resident who is caught bootlegging. The consequence is that booze has practically been eradicated from the camp. When the town was real young it had a gambling den called the "Red Apple." Roulette and faro were part of all camps and Picher was no exception. But prospectors and single men gradually were displaced by family men and the "Red Apple" has gone and not even the core is left. Houses, hotels, and rooming houses are full and a building boom has started to provide facilities for a fast increasing population. The weekly payroll from the mines alone is estimated at $180,000. The

> miners are all American and free from the influences of labor organization. . . . Its forced growth has necessarily been of a shack character. There it sprawls on the open plain, a gangling, awkward, disheveled caricharacter, hot as an inferno under its noon sun, and treeless as a desert. Up and down its main street the traffic of the district swings ceaselessly; taxis sweating in from Miami and Joplin or returning; motor trucks creaking under their loads of machinery; ore wagons lumbering slowly past; and clouds of dust swirling forever. Sanitary conditions are of course elementary. There is no public utility service. The water supply comes from deliveries, but its method of storage is the most primitive barrel which stands in the front yard. Mostly those with its top off, willingly receptive to whatever wandering germ or strolling bacillus comes that way.[12]

The final lines of the *Daily Oklahoman*'s characterization of Picher suggest one of the Tri-State District's most pervasive social problems—inadequate housing for the miners. From the beginning the workers seemed to be in an eternal hurry. They rushed in to prospect so as to strike ore as soon as possible. Promoters hastened to develop the region quickly. When miners ceased being operators and became day workers, they were in a hurry to get to work so as to earn the good wages paid, especially if they were shovelers. One observer called this condition the Joplin Colic.[13] Probably then, in the beginning the great neglect of housing was due to haste. This does not explain, however, the perpetuation of this condition. The land system undoubtedly made a contribution. Land and royalty companies controlled most of the mining land, either in fee or lease.

Lack of transportation facilities made it necessary for the miner to live as close as possible to his work. The land was divided into mining plots and leased to miners. They could also lease lots for home building purposes. At Granby, for example, the miners were allowed to build dwellings and fence a garden plot free of charge on company land.[14] The Rex Mining Company permitted miner lessees to build homes near their mining leases on its Joplin Thousand Acre Tract.[15] Needless to say, these homes were poorly constructed. In their rush to get prospects under way, the miners gave little attention to comfort and sanitation. The important thing was to have minimum shelter available. Many of the miners were from the nearby Ozark hill country, and were accustomed to little better than a log shanty. Also, since most early dwellings were on mining land and largely undermined, there was less incentive to build a better home. The chief interest was to locate as close to the diggings as possible.[16]

While miner housing in the district was substandard, there is no evidence that any sort of company town, complete with company store and scrip, common in the coal fields, ever developed in the Tri-State District. The closest to this was the policy of the Missouri Lead and Zinc Company, an enterprise holding 1,300 acres of land near Joplin. This company divided its land into mining plots, laid out a residential district, and erected 350 miners' homes. A self-sufficient community, complete with company lumber yard, blacksmith shop, and store, were added to meet the miners' needs. But whether mining on their own or working for wages, the miners were paid in cash.[17] In modern times, Picher, Oklahoma, most approximates a company town. In 1951, because the town was extensively undermined, a part of the business district was in danger of caving. The Eagle Picher Company granted financial assistance to businessmen in moving buildings and improvements to a safe location.[18]

The Missouri Bureau of Labor Statistics noted substandard housing in 1889 as one of the big problems in the district. The report lamented that this was needlessly so since wages were good. "The trouble is," concluded the bureau, that "the savings which could certainly buy a comfortable home are dumped into holes in the ground through prospecting for new deposits."[19]

Anthony J. Lanza made a house-to-house visit around Webb City in 1914 exploring housing conditions. He noted that

> generally speaking, among miners in Webb City, home conditions are fair to good, whereas those living in the outskirts of town, or on mining land between towns, were bad. Taken all in all when wages of the miners in Southwest Missouri are considered, home conditions are far below par as far as sanitation and comfort are concerned. The situation in this respect is remarkable, because it is so needlessly bad. The miners make $3.50 to $5.00 per day, and even more at times, and they do not migrate to the extent observed in other mining communities. The chief obstacles in the way of improvement are a failure to appreciate better living conditions, and possibly to a lesser extent, the fact that many families live on mining land upon which nothing but temporary shacks can be built.[20]

Lanza visited a total of 694 homes in his survey and noted that most of them were one, two, and three-room "shacks." On the state of cleanliness in these dwellings he recorded that 317 were good, 318 fair, and 59 were classified as bad.[21] Lanza found the water and sanitation facilities objectionable too.

> The water supply of a great number of homes is rather unique. Water of good quality is obtained from deep wells and is peddled around the district in water wagons and sold by the barrel wherever there are no water pipes. Wells are scarce, and in the majority of the homes the water barrel suffices for cooking and personal needs. . . . The outhouses are wretched, a feature in which southwestern Missouri resembles a great part of the rural communities of the United States. In 680 premises there were 644 insanitary privies, which consisted of the simplest kind of a box struc-

> ture over a shallow pit dug in the ground. There were sewer connections in 36 premises. In view of the prevalence of filthy privies all over that part of the country, the scarcity of wells is fortunate, and undoubtedly the fact that water is peddled from a pure source is the greatest factor in preventing widespread typhoid fever and other intestinal disorders. . . . In none of these homes was there a bathtub or bathing facilities other than could be obtained from a pan of water on the kitchen floor. In 281 premises there were 92 cases of tuberculosis and 120 cases of miners' consumption.[22]

Until 1940, the outside privy was a common sight in some portions of the district. While most of the dwellings had the benefit of city water, some wells and many cisterns were used on the periphery of district towns. From the Department of Labor Conference on Working and Housing Conditions, held at Joplin in 1940, it was noted that miner housing was still largely substandard.[23] Investigators traveling around the fringe of Joplin, Galena, and Picher reported the presence of "mining slums."

The integration evident in the Tri-State District's economic and social life ramified into its political, religious, and intellectual life too, though with some exceptions. The political complexion of the district was by no means a one-party section. In thirteen elections the district witnessed a political split. On the other hand, in sixteen elections the Tri-State District voted as a unit, ten times Democratic, six times Republican. Of the fifteen presidential elections between 1896 and 1950, the district voted with the national majority eight times: for Theodore Roosevelt in 1904; in 1912 and 1916 for Woodrow Wilson; 1920 for Warren Harding; 1928 for Herbert Hoover; 1932 and 1936 for Franklin Roosevelt; and 1952 for Dwight Eisenhower.[24]

In terms of religious life, the Tri-State District followed rather closely the rule of integration. The region was dominantly Protestant, with a wide range of faiths represented. Also, it was evident from the comparative statistics that Tri-State counties were much less interested in religious affiliation than their population counterparts. This undoubtedly reflected the rough, boisterous, and oftentimes reckless attitudes characteristic of mining camps. What the statistics did not show is the remarkable independence of doctrinal view and fundamentalism exhibited by local congregations, even those associated with national denomination groups with a reputation of liberalism in doctrine and antifundamentalism. Among the Protestants, the Baptists were the most numerous, followed in order by the Disciples of Christ, Methodists, and Presbyterians.[25]

Tri-State intellectual life reflected in many respects the traditional mining camp contempt for refinement and amenities. The mining counties showed, if not an indifference, at least a retardedness in providing adequate educational facilities. Some exceptions were found, however, to this general rule. Mabel Draper, early Joplin pioneer, recounted that at times the miners showed a strong enthusiasm for educating their children. According to her, a Joplin town meeting in 1872 resulted in the organization of a school board for the purpose of erecting a two-room school house. She added, "The money came in a hurry, one thousand dollars by popular subscription, and a few days later, right up the hill from us there were piles of rock from the nearby mines for the foundation and stacks of good smelling new lumber."[26]

By 1876, Joplin had two grammar schools, while Carthage boasted a $36,000 school building with twelve teachers.[27] By 1950, Miami, Neosho, Carthage, Galena, Webb City, and Joplin had public libraries which contributed to the general intellectual uplift of the district. These larger communities had at least average educational facilities. The smaller mining communities, however, pulled the district average down. This was reflected in the district expenditure for each pupil for education. Not only were the Tri-State counties low compared to other counties in their respective states, but they were likewise considerably lower than their respective average state per capita expenditures. For example, in 1950, Cherokee County spent $240.98 for each pupil, while the state of Kansas spent $291.31; Jasper County spent $169.92 compared to Missouri's average of $198.24 for each pupil; and Oklahoma spent $149.75 per capita compared to Ottawa County's $133.97.[28] If the Tri-State District was marginal in expenditures for education, it also manifested an unusually heavy pupil loss in the upper age group.[29]

The Tri-State cultural heritage is saturated with the concept of individualism, a noteworthy trait best illustrated by the "poor man's camp" tradition. To the layman, this means that a workman could engage in mining operations with only a small amount of capital, and if his prospect were fortunate, he stood a good chance of becoming wealthy. James Bruce, a local mining engineer, noted the wide opportunities there as late as 1912 when he wrote, "The Tri-State District is a poor man's camp and almost every miner who has spent any number of years in it . . . has at some time owned a prospect . . . and has made at least some attempt to organize a company, to secure a lease, and try his luck at finding diggings."[30]

The sociologist has a term, mobility, that is useful in characterizing this facet of Tri-State social life. Because of the "poor man's camp" tradition, there has been considerable vertical mobility. That is, according to the Missouri labor commissioner in 1889, men working for a daily wage in the mines customarily saved a portion of their earnings to grubstake them in prospecting:

> There are few large companies employing men by the week or month. The rule is for each man to mine for himself, paying a stipulated royalty to the owner of the land. If within a few days the prospector finds nothing, he takes his rope and windlass and bucket to another spot of land, digs

> another hole and prospects again, keeping on thus until ore is discovered. Miners who work for the few companies in the country save their wages, and when a hundred or so of dollars have been accumulated, sink it into some prospecting shaft. One miner is now as poor as when he began, but is rich in anticipations. He will save his wages, expecting that at last the next prospecting will prove a bonanza. The successful few are heard of not once, but time and time again. The story of how one man digging a well for water in his back yard found, instead of water, a vein of ore from which he took out $100,000, is told around almost a hundred thousand times. On the other hand, the thousand who spend their surplus wages in prospecting for lead, but who never find it, are rarely heard of. One man reported to the Bureau that he sunk two years' savings looking for ore, then returned to work for more money, but before saving enough to continue his prospecting, another man jumped his claim and going six inches deeper struck an exceedingly rich vein. Owing to this gambling spirit but few miners seem prosperous or own their own homes, although wages are fair.[31]

The scattered nature of the deposits, and their relatively shallow depth enabled miners with limited funds to seek ore. The thousands of shallow test pits around Joplin attest to the hope and labor of miners to become wealthy. Wiley Britton, an early resident of the district, estimated that "probably one out of fifty to one-hundred would not be too low an estimate" for those who struck it rich.[32]

Those who made a rich strike moved from the wage earner to the operator class, and just as there was social movement upward in the social scale, based on economic success, so was there a similar movement back to the wage earner class if the prospect did not produce. Unwise investments in new prospecting or careless spending proved the undoing of many a successful miner.[33]

The word *miner* is anomalous in the Tri-State District and

produces a problem in semantics for the researcher. It meant both one who worked for wages as well as an operator. The operator worked too in the early days, and the names of operator associations reflect this, since there was the Southwest Missouri and Southeast Kansas Zinc Miners Association. This group consisted of mine owners seeking through concerted action better markets and investors in district mines.[34] Only after 1900 can one discern a more specialized use of the word *miner*. Thereafter it meant an employee working for a daily wage, and *operator* came to be used as a designation for the mine owner. In 1889 the Missouri labor commissioner made a study of district economic conditions, and he found the gambling spirit of local miners toward prospecting, "dumping their savings into holes in the ground," a major cause of hardship and social problems.[35]

While there was vertical mobility in the district, there was a general lack of mobility in terms of space until after 1920. That is, between 1860 and 1920, once people came into the district, they generally remained. Between 1860 and 1920, the Tri-State District not only held its own in terms of population numbers, but actually showed an increase for the inclusive decennial periods.[36]

In spite of the fact that the Tri-State District showed no loss of population as a region until 1920, there was considerable mobility in local mining towns. Strikes in the Galena field drew miners from Joplin, and the Picher strike attracted people from all over the district. Anthony Lanza noted this intradistrict migration in 1915, but he was impressed by the regional population stability. "In most mining camps there is considerable annual migration, but in the Joplin District, the miners are natives and previous to 1915, outsiders had not come in any large number."[37]

Intradistrict migration was lamented by the Granby *Miner* in 1873 as responsible for nearly depopulating Granby since "many miners, naturally migratory, were deceived by the blowing of newspapers and went to Joplin."[38] The Missouri labor commissioner commented in 1889 that "those who have lived here for several years and made it their home seem well contented, but the restless ones spend most of their time prospecting, become dissatisfied with the work, and go away for awhile, though they generally come back and commence prospecting again."[39]

That the population remained stable in the district through the years was due to the fact that as long as there were jobs in the mines, workers and their families had the means of support. Even in periods of national prosperity, though, the mines might be shut down for short periods mainly on account of sagging ore prices or a surplus of concentrates.[40]

Such a situation in the mines of Arizona or Colorado would have set off a migration of workers to other sections. In the Tri-State District it was common, until around 1920, for the more thrifty workers to have a small farm or garden plot and a cow, a pig or two, and chickens. The wives and children generally took care of the livestock, and the miners farmed or gardened after work and in periods of unemployment. The women canned food, and the pigs were butchered and processed for winter's use. Until recently, the wooded sections around the mining camps furnished abundant wildlife for food as well as fuel for cooking and heating. The many district streams, unless contaminated by mine water, supplied bass and channel catfish for the family larder.

No other mining region in the country has been more capable of sustaining its people than the Tri-State District. Until recent times, even during economic distress, a miner could support himself and his family. Of course those miner families living in camps like Picher, where the mines and tailing piles extended literally to their residences, no such economic independence was possible, and in times of mine distress these people suffered considerably.

Another reason for the population remaining fairly stable

in the past is that, whenever a depression set in, unemployed miners would lease ground in the shallow deposit field and prospect. Most of them earned enough to support their families by gouging for ore with crude methods and equipment. In 1914 the *Engineering and Mining Journal* noted this tendency in attempting to explain "Why Joplin Does Not Languish":

> When a miner is pushed out of work by a shutdown, he often helps develop new fields. He becomes a prospector and producer . . . when he loses his job. Often, miners are found in old diggins' working surface gouges and shallow lead deposits. Only a small outlay is required in capital . . . a hand windlass, a barrel sawed in half for tubs, a rope, a few strong hands, a few picks and shovels, and enterprise enough to rig up and dig a hole to the ore. He is generally able to get enough timber around the ground to crib up the shafts and provide sufficient timbering to hold the ground long enough to accomplish the work of getting the ore.[41]

During hard times some miners were not too concerned with the formality of obtaining a valid lease from the land owner before working a mine. Some miners, according to the *Engineering and Mining Journal*, even pirated ore from operating mines that were shut down temporarily: "It seemed like old times last week when petty ore thieving was reported from Webb City, Missouri. A few miners were operating small mines in the older Missouri camps or working large mines on a small scale. They were mostly gouging for lead ore and it is these that suffered from the thieving. All junk dealers of the district were notified and they will have a hard time disposing of it. In former times there was much of this but it has largely disappeared during the past two years."[42] Even during the Great Depression following 1929 some of the more resourceful miners of the district, reported the *Journal*, sought to make some sort of a living by prospecting: "Back to the gouges is the slogan of the Tri-State District. There are 3,000 miners out of work. About 1,000 are operating small prospects over the district. Many are anxious to prospect for shallow deposits but lack the capital. Hand windlasses, horse hoisters, buck rocks, hand jigs, and sluice boxes again are in vogue and this is furnishing a livelihood for many district miners."[43]

The fact that many of the workers were recruited from the farms of the district furnishes another explanation of population stability. Malcolm Ross wrote facetiously that the "Ozark hills are rich in hungry hillbillies," and he recounts that one operator claimed that "all you have to do to get fresh miners is to go out in the woods and blow a cowhorn."[44] Whenever the mines were open, farmers' sons would go to the mines, and when operations ceased, they returned to the rural areas.

Also, as the mines became more highly mechanized, requiring less workmen, many miners were absorbed in other industries recently established in the district. These included a nitrate plant and wood processing, roofing manufacturing, dairy processing, electronics manufacturing, and automobile tire production concerns. Considerable employment was furnished by the foundaries and shops manufacturing mine and milling equipment. Implement, seed, and livestock feed establishments served the district farm needs and supplied some employment. Three powder companies, in addition to supplying mine explosive needs, also had defense contracts which provided additional jobs. A large army training center at Neosho, established in 1942, absorbed many displaced miners too.

Joplin's position as a distribution point for regional needs in food, household furnishings, and other commodities produced employment for many former miners. Until recently it was possible to escape the problems of employment from mining in a fashion noted above. Recently, however, it has been not as easy to gain security on an individual basis. One of the reasons for the slowness of the labor movement in taking hold in the district was because the miner could take care of himself one way or another. Unionization developed in proportion to the miner's growing inability to meet his own needs.

CHAPTER XVI

The Tri-State District at Mid-Century

From 1850 to 1950 the Tri-State District was the world's leading producer of lead and zinc concentrates, accounting for 50 per cent of the United States' zinc production and 10 per cent of its lead production. The value of this century of production is in excess of one billion dollars. These Tri-State mines made a substantial contribution toward equipping the United States for four major wars, in addition to contributing to the high standard of living enjoyed by the American people.[1]

Industry in the United States and abroad drew heavily on Tri-State lead and zinc concentrates to produce, besides arms and munitions, bearings, castings, electrical fittings, pipe, galvanized metals, tubing, wire, brass, die castings, bronze, battery zinc, hat blocks, toys, coins, chains, nails, screens, bolts and hardware, airplane and automobile equipment, storage batteries, fuse plugs, cooking utensils, landing gear, pipe organs, laundry equipment, photo engraving parts, roofing, surgical equipment, enamels, interior paints, linoleum, rubber, white shoe polish, printing inks, chemicals, matches, cosmetics, pharmaceutical goods, ceramics, automobile tires, dental cement, fireworks, aniline dyes, hydrogen gas, sugar purification, cyanide process for refining gold and silver, salts and pigments, calico printing, drugs, fertilizer, oil refining, sewage disposal, celluloid, starch, syrup, china ware, die lubricant, pipe fittings, pottery, sanitary equipment, varnishes, electric light bulbs, optic glasses, insecticides, telephone cables, electric cables, gaskets, x-ray apparatus, babbit metals, solders, type metals, metal packing, atomic research, lead foil, radium casing, clock weights, piano tubing, and a host of other products and uses.

By 1950 the Tri-State District as a mineral-producing area was largely marginal; that is, the rich ores had been extracted, and there remained only low-grade minerals. United States Bureau of Mines engineers estimate that the region contains a reserve of 66,100,000 tons of crude ore, sufficient to mill out 2,402,433 tons of zinc concentrates and 312,649 tons of lead concentrates. In terms of industrial use, this is sufficient zinc for 20,000,000 automobiles, 650,000,000 sheets of galvanized roofing, 11,800,000 miles of barbed wire, lead for 11,000,000 storage batteries, and 18,500,000 gallons of paint. At the Eagle Picher Company's Central Mill at Commerce, Oklahoma, new milling practices derived small quantities of germanium and cadmium from lead and zinc crude ore. In the ore reserve estimates it is believed that there are sufficient quantities of germanium to produce 16,000,000 radar rectifiers and of cadmium to produce 7,500,000 pounds, used principally for high-speed bearings and plating on automobile grills.[2]

In order to mine the marginal ores at a profit, Tri-State operators required some sort of subsidy. The most popular type of governmental assistance has been increased tariff protection. The high postwar concentrate price for each ton, $246.50 for lead and $135.00 for zinc in 1951, began to decline rapidly early in 1952. This was due largely to the policy of the federal government following the outbreak of the Korean War. In order to stockpile strategic metals, the Federal Defense Minerals Procurement Agency signed contracts with Peruvian, Mexican, and other foreign lead and zinc producers, guaranteeing them six cents a pound above the prevailing United States market price for lead and zinc concentrates.[3] This policy had far-reaching repercussions in the domestic lead and zinc producing fields, especially in the Tri-State District because of its low-grade ore. By the middle of 1953, district operations were paralyzed by the foreign imports. The

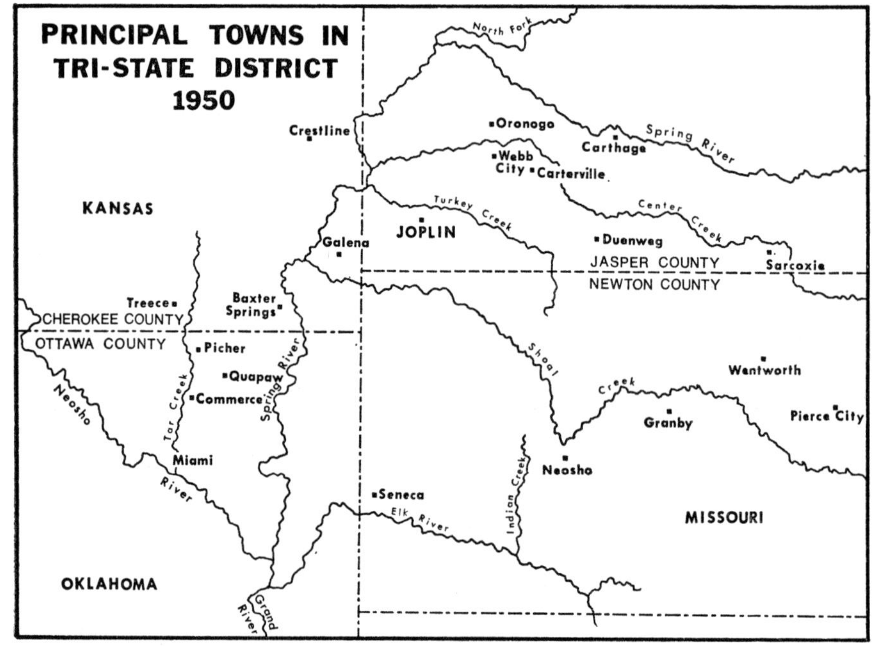

high price for concentrates in 1951, $246.50 for lead and $135.00 for zinc, had declined to $154.74 a ton for lead and $65.00 for zinc.[4]

Editorially, regional newspapers attacked the federal government for its favoritism shown to foreign producers. The *Daily Oklahoman* was especially sharp in its criticism, claiming that the federal government was "Serving America Last":

> At the outbreak of the Korean war government purchasing agents went traipsing off to the earth's far corners to put strategic metals under contract for future delivery at fixed prices. Less than three years later the government has on its hands a mounting metal surplus, much of it coming from foreign sources at prices above the domestic level. But more than a loss to the taxpayers was involved in these

> purchasing commitments based on the condition of the metals market three years ago. What happened also was that domestic mining was curtailed and workers thrown out of jobs by foreign production enjoying a subsidy from the American taxpayers. One effect close to home is the curtailment of mining operations in the Tri-State zinc and lead producing area of northeast Oklahoma. From a peak of 6,000 jobs during World War II employment in that area has slipped to about 1,800.[5]

The Joplin *Globe*, long the local spokesman for the district's lead and zinc industry added:

> Representative Ed Edmondson of the Second Congressional District of Oklahoma will visit Miami April 10 to discuss with mine operators price and production problems of lead and zinc in the Tri-State field. He is thinking in terms of possible legislative relief. Maintenance of lead and zinc prices at a level calculated to encourage continued mining activity in the Tri-State field is not of course a local problem. It has national and international aspects. The decline of lead and zinc prices has been caused by the accumulation of zinc and lead surpluses. Some operators have suggested a sliding-scale of protective tariffs. They believe it is plain economic stupidity to give away or loan large sums of taxpayer money to foreign countries to finance and stimulate production of metals, among other things, only to have the excess supply depress the American market and jeopardize taxpayer jobs.[6]

Representative Ed Edmondson, of the Second Congressional District of Oklahoma, conducted hearings in Miami, Oklahoma, on April 10, 1953, with operator and labor representatives in attendance. After a week of local investigations Edmondson reported to the House of Representatives that

> conditions in the Tri-State District must be rated as crucial. . . . In our district between 80 and 90 of the smaller operators have been forced to close down, many of them broke.

> Most of the larger companies are operating at a loss, trying to run until Washington will do something to give zinc mines a fair deal. The Tri-State district here has a population of about 85,000 people. Nearly every person has been hurt directly or indirectly by this unreasonable ore price. We estimate that 50 percent of the men who were working in the mines are out of work and the ones that are working have taken a cut in wages of over $100 per month. When foreign labor is causing this, something can be done and something should be done.... The exodus has cut the population of Picher from about 5,000 to between an estimated 3,500 and 4,000.[7]

The United States House of Representatives, responding to Edmondson's request and to other representatives of lead- and zinc-producing areas, authorized the establishment of the Select Committee on Small Business, with Representative William Hill of Colorado as chairman. The committee conducted hearings throughout the lead and zinc fields of the United States, and representatives of operators as well as labor appeared before the group. In response to the findings of the Hill Committee,[8] two measures looking to the relief of the United States lead and zinc industry were proposed in Congress during 1953. The Simpson Bill sought to place a tariff on lead and zinc imports on a sliding scale basis so as to make the price of foreign-produced concentrates higher than domestically produced concentrates. The Johnson Bill proposed that a Metals Commodity Corporation be established. This agency would guarantee a price for each ton higher for domestic lead and zinc than was paid by the government to foreign producers under the Defense Procurement Program. Both proposals were defeated in July, 1953.[9]

Meanwhile, congressmen representing lead- and zinc-producing areas continued to promote relief for the metal industry, all in the face of mounting foreign imports of lead and zinc concentrates. Representative George Dawson of

Utah announced before the House of Representatives on January 25, 1954, that

> in 1951, 9 of the 11 Western States produced 635,000 tons of lead and zinc, 60 per cent of the Nation's total production, worth $206 million. Nineteen hundred and fifty-three production dropped to 482,000 tons of lead and zinc worth $100 million. This was less than half the worth of 1951 production. The decreased worth reflects both production loss and decreased prices. . . . The dollars spent for domestic mine production are effective to their full value in our economic blood stream. It is questionable that more than a small portion of the dollar spent for foreign metals returns to circulation here. The few cents per pound saved by buying foreign metal, say by the automobile industry, undoubtedly will cost them more than the savings made, through inability to sell cars in the domestic areas suffering the loss of income. Once closed down, mines would take years and tremendous sums to reopen. This Nation could not disarm more effectively than to lose its power to produce metals in an emergency.[10]

Representative Edmondson followed with an appeal for congressional relief for the Tri-State District and the national metal industry.

> It has been a concern of mine ever since coming to this House last January to witness the deterioration and the destruction of the lead and zinc mining industry of the Tri-State area. There can be no question about the fact that this is a direct result of the mounting imports which have greatly increased during the past several years. Nineteen fifty-two broke all records for imports of zinc. Nineteen fifty-three, the first 6 months, showed a 245 percent increase over the first 6 months of 1952. So imports continue to mount while more and more men become unemployed in this industry. . . . Imports of lead and zinc metals produced at a much lower cost than the American markets show have depressed the market to the extent that most operators cannot make a

> profit on the mining operation. . . . Production of lead and zinc concentrates in the Tri-State District has dropped materially during this same period of 1952–1953. The production of zinc concentrates in the early part of 1952 averaged about 15,000 tons a month in this district as compared to about 4,000 tons a month at the present time. . . . The situation is serious. The President of the United States has acknowledged the seriousness of the situation. The Secretary of the Interior has acknowledged the seriousness of the situation. We will not have a gun to fire any kind of ammunition at the enemy if we do not maintain our lead and zinc supply in this country. It seems to me imperative that something should be done about it.[11]

The Tri-State District was approaching the end of an era. Even with relief to the lead and zinc industry sufficient to make the mining of the remaining ore reserves profitable, it was estimated that these reserves would be exhausted in perhaps five years, and at most there remained only about ten years of mining.[12]

Mining and milling of lead and zinc ores declined in the Tri-State District during the 1950's, and in the next decade operations ceased. Lead and zinc mining had been the principal means of support for the region for a century. The lack of a sufficient number of industrial jobs to absorb the unemployed miners caused a migration of many Tri-State residents to more attractive parts of the country. Picher, Oklahoma, and Galena, Kansas, were nearly depopulated by the miner exodus. The Ottawa County population in 1960, at 28,301, had declined 12 per cent during the decade of the 1950's. Jasper County, for the same period, with a population of 78,863, registered about a 1 per cent decline. The migration caused by the closing of Tri-State mines and mills reduced the population of Cherokee County, Kansas, by 12 per cent to 22,279.

Jasper County business and civic leaders, centering on Joplin, and Ottawa County business and civic leaders, centering on Miami, energetically recruited new job-producing industries to stem the tide of migration. Their efforts produced an economic metamorphosis in the Tri-State District. The analogy from agriculture, moving from the one-crop system to diversified farming, is applicable to the Tri-State District in its new age of industrial regeneration. Electric and water-power development, transportation expansion, and communication extension occurred to sustain new industries at Joplin, Miami, and other district towns. New industries include firms producing electronic devices; photographic equipment; metal castings; chemicals; motor parts, including intricate bearings; and textiles. The natural beauty of the region has been developed as tourism resources to provide camping, hunting, water sports, and general out-of-doors recreation. A concomitant expansion in agriculture, dairying, and stock raising have made the Tri-State District a national leader in food processing. Signs of success in the economic rehabilitation for the Tri-State District are found in the 1970 census returns. Ottawa County had nearly a 30,000 population, a 5 per cent increase over 1960. Jasper County, with nearly 80,000, registered nearly a 2 per cent growth in population. Only Cherokee County, with 21,549, showed a decline, 4 per cent, in population.

Notes

I. THE REGION AND ITS MINERAL DEVELOPMENT

1. Otto Ruhl, Simeon Allen, and Stephen Hold, *Zinc-Lead Ore Reserves of the Tri-State District, Missouri-Kansas-Oklahoma,* U.S. Bureau of Mines *Report of Investigations* No. 4490, 11.

2. W. S. Tangier Smith and C. E. Siebenthal, *Geological Atlas of the United States: Joplin District Folio,* U.S. Geological Survey *Report* No. 148, 1.

3. U.S. Department of Agriculture, *Yearbook of Agriculture, 1941: Climate and Man,* 873, 945, 1067.

4. A. J. Martin, *Summarized Statistics of Production of Lead and Zinc in the Tri-State Mining District,* U.S. Bureau of Mines *Information Circular* No. 7383, 5.

5. U.S. Bureau of the Census, *Seventeenth Census of the United States: 1950. Population,* I, 328. Recent population statistics for representative district towns show Galena, Kansas, 4,029; Baxter Springs, Kansas, 4,647; Commerce, Oklahoma, 2,442; Miami, Oklahoma, 11,801; Quapaw, Oklahoma, 938; Picher, Oklahoma, 3,951; Joplin, Missouri, 38,711; Neosho, Missouri, 5,790; Webb City, Missouri, 6,919; Carterville, Missouri, 1,552; Oronogo, Missouri, 519; Granby, Missouri, 1,670; and Waco, Missouri, 177.

6. Clarence A. Wright, *Mining and Treatment of Lead and Zinc Ores in the Joplin District, Missouri, A Preliminary Report,* U.S. Bureau of Mines *Technical Paper* No. 41; George M. Fowler and Joseph P. Lyden, "The Miami-Picher Zinc-Lead District," *Economic Geology,* Vol. XXIX (June, 1934), 390–96; "The Ore Deposits of the Tri-State District," *Economic Geology,* Vol. XXVIII (January, 1933), 75–81; Charles F. Jackson et al., *Lead and Zinc Mining and Milling in the United States,* U.S. Bureau of Mines *Bulletin* No. 381; C. E. Siebenthal, *Origin of the Zinc and Lead Deposits of the Joplin Region,* U.S. Geological Survey *Bulletin* No. 606; H. F. Bain, C. R. Van Hise, and G. I. Adams, *Preliminary Report on the Lead and Zinc Deposits of the Ozark Region,* U.S. Geological Survey, *Twenty-second Annual Report;* and Samuel Weidman, C. F. Williams, and Carl O. Anderson, *Miami-Picher Zinc-Lead District.*

7. Wallace H. Witcombe, *All About Mining,* 155.

8. Bain, Van Hise, and Adams, *Preliminary Report,* 215.

9. *Ibid.*, 151.
10. W. S. Tangier Smith, *Lead and Zinc Deposits of the Joplin District*, U.S. Geological Survey *Bulletin* No. 213, 200.
11. Wright, *Mining and Treatment*, 10.
12. Bain, Van Hise, and Adams, *Preliminary Report*, 204–205.
13. W. S. Tangier Smith, *Water Resources of the Joplin District*, U.S. Geological Survey *Paper* No. 145, 79.
14. Smith and Siebenthal, *Geological Atlas*, 14.
15. Weidman, Williams, and Anderson, *Miami-Picher District*, 77. Also see Carey Groneis and William Krumbein, *Down to Earth, an Introduction to Geology*; William H. Emmons et al., *Geology, Principles and Processes*; and Edward S. Dana, *A Textbook of Mineralogy*.
16. Erasmus Haworth, W. R. Crane, and A. F. Rogers, *Special Report on Lead and Zinc*, Kansas Geological Survey *Report* No. 8, 448.
17. Joplin Chamber of Commerce, "Joplin Mineral Resources" (Joplin, n.d.), 2. Mimeographed.
18. Wiley Britton, *Pioneer Life in Southwestern Missouri*, 25–26.
19. Weidman, Williams, and Anderson, *Miami-Picher District*, 23.
20. Wright, *Mining and Treatment*, 7.
21. *Mineral Production of Oklahoma from 1901*, Oklahoma Geological Survey *Bulletin* No. 15, 43.
22. Weidman, Williams, and Anderson, *Miami-Picher District*, 57.
23. *Mineral Production from 1901*, 43.
24. Weidman, Williams, and Anderson, *Miami-Picher District*, 58.
25. Clarence A. Wright, *Mining and Milling of Lead and Zinc Ores in the Missouri-Kansas-Oklahoma Zinc District*, U.S. Bureau of Mines *Bulletin* No. 154, 7, 11, 16.
26. Weidman, Williams, and Anderson, *Miami-Picher District*, 62.
27. Walter Williams, *The State of Missouri*, 169, and Witcombe, *All About Mining*, 154.
28. Erasmus Haworth, *Annual Bulletin on Mineral Resources of Kansas for 1897*, 21.
29. Wright, *Mining and Milling*, 9–10.
30. Haworth, *Annual Bulletin, 1897*, 20, and Witcombe, *All About Mining*, 147.
31. Williams, *State of Missouri*, 169.
32. Haworth, *Annual Bulletin, 1897*, 19.
33. Jackson et al., *Lead and Zinc Mining*, 19–20.
34. Haworth, *Annual Bulletin, 1897*, 27.
35. Jackson et al., *Lead and Zinc Mining*, 22.
36. Doss Brittain, "Ground Breaking in the Joplin District," *Engineering and Mining Journal*, Vol. LXXXIV (August 10, 1907), 255. *Engineering and Mining Journal* is hereafter cited as *EMJ*.
37. James R. Finlay, *The Cost of Mining*, 313.
38. Oliver W. Keener, *Method and Costs at Hartley-Grantham Mine*, U.S. Bureau of Mines *Information Circular* No. 6286, 2.

39. Temple Chapman, "Decline of Sheet Ore Mining at Joplin," *EMJ*, Vol. XCI (May 6, 1911), 911.

40. Brittain, "Ground Breaking in the Joplin District," *EMJ*, Vol. LXXIV (August 10, 1907), 255.

41. Wright, *Mining and Milling*, 10.

42. Jackson et al., *Lead and Zinc Mining*, 22.

43. Chapman, "Decline of Sheet Ore Mining," *EMJ*, Vol. XCI (May 6, 1911), 911.

44. Haworth, *Annual Bulletin, 1897*, 27.

45. Weidman, Williams, and Anderson, *Miami-Picher District*, 27.

46. Brittain, "Ground Breaking in the Joplin District," *EMJ*, Vol. LXXXIV (August 10, 1907), 61.

47. Keener, *Method and Costs at Hartley-Grantham Mine*, 1–2.

48. Brittain, "Ground Breaking in the Joplin District," *EMJ*, Vol. LXXXIV (August 10, 1907), 61.

49. Chapman, "Decline of Sheet Ore Mining," *EMJ*, Vol. XCI (May 6, 1911), 911. Also see Finlay, *Cost of Mining*, 313.

50. Weidman, Williams, and Anderson, *Miami-Picher District*, 69.

51. Witcombe, *All About Mining*, 148; also see Keener, *Method and Cost*, 1–2.

II. DISTRICT DEVELOPMENT TO 1865

1. The Crozat Patent, *American State Papers, Public Lands*, VIII, 575. Also see Walter Williams and Lloyd C. Shoemaker, *Missouri, Mother of the West*, I, 90–91.

2. Arthur Winslow, *Lead and Zinc Deposits*, 267–71.

3. Paul M. Tyler, *From the Ground Up: Facts and Figures of the Mineral Industries of the United States*, 140.

4. Williams and Shoemaker, *Missouri*, I, 90–91.

5. Louis Houck (ed.), *The Spanish Regime in Missouri*, I, 315.

6. Henry R. Schoolcraft, *A View of the Lead Mines in Missouri*, 14–19. For mining in early Virginia, see Witcombe, *All About Mining*, 145.

7. Moses Austin to the President of the United States Concerning Lead Mines in Upper Louisiana, Mine a Burton, February 13, 1804, *American State Papers, Public Lands*, I, 188.

8. Edward Tiffin to the Congress of the United States on the Lead Mines of the Missouri Region, *American State Papers, Public Lands*, II, 658–714.

9. Moses Austin to the President, *American State Papers, Public Lands*, I, 188.

10. Amos Stoddard to the President of the United States Containing a Description of the Lead Mines of Upper Louisiana, *American State Papers, Public Lands*, I, 188.

11. Schoolcraft, *A View*, 254.

12. Stephen H. Long, Reuben G. Thwaites (ed.), *Stephen H. Long*

Expedition, 1819–1820, in *Early Western Travels, 1748–1846*, XVII, 230–323.

13. Henry R. Schoolcraft, *Journal of a Tour into the Interior of Missouri and Arkansas*, 54.

14. John R. Holibaugh, "Early Mining in the Joplin District," *EMJ*, Vol. LVIII (December 1, 1894), 508.

15. *Illustrated Historical Atlas of Jasper County, Missouri*, 14.

16. *Ibid.*, 19. Also see F. A. North, *The History of Jasper County, Missouri*, 153.

17. Joel T. Livingston, *A History of Jasper County and Its People*, I, 141.

18. North, *Jasper County*, 390, and *Historical Atlas*, 19.

19. John R. Holibaugh, *The Lead and Zinc Mining Industry of Southwest Missouri and Southeast Kansas*, 32.

20. Britton, *Pioneer Life*, 17.

21. Joplin *Globe*, January 31, 1943.

22. "History of Eagle Picher," *EMJ*, Vol. CXLIV (November, 1943), 70.

23. *Historical Atlas*, 15, and Joplin Chamber of Commerce, "Joplin Missouri: History and Development" (Joplin, 1933), 2. Mimeographed.

24. "Mine Owners of Joplin," *EMJ*, Vol. L (July 26, 1890), 96.

25. Dolph Shaner, *The Story of Joplin*, 1.

26. Winslow, *Lead and Zinc Deposits*, 447–50.

27. Moses Austin to the President, *American State Papers, Public Lands*, I, 188.

28. George C. Swallow, *The First and Second Annual Reports of the Geological Survey of Missouri*, 163–64.

29. Britton, *Pioneer Life*, 20.

30. Holibaugh, "Early Mining in the Joplin District," *EMJ*, Vol. LVIII (December 1, 1894), 508.

31. *Historical Atlas*, 22.

32. Livingston, *Jasper County*, I, 34, and North, *Jasper County*, 487–88.

33. George C. Swallow, *Geological Report of the Country Along . . . the Pacific Railroad*, 37.

34. *Historical Atlas*, 22.

35. Swallow, *First and Second Annual Reports*, 164.

36. Granby *News Herald*, Centennial Edition, June 1, 1950.

37. *Ibid.*, and Whitcombe, *All About Mining*, 156.

38. Swallow, *First and Second Annual Reports*, 160.

39. *Ibid.*, 161, and Swallow, *Geological Report*, 36.

40. Winslow, *Lead and Zinc Deposits*, 286.

41. Granby *News Herald*, Centennial Edition, June, 1950.

42. Swallow, *Geological Report of the Country*, 36.

43. W. F. Switzler, *Illustrated History of Missouri from 1554 to 1881*, II, 566.

44. William S. Moseley, "Paper on the Lead Mines of the Southwest," *Western Journal and Civilian*, Vol. IV (1851), 412.

45. Holibaugh, "Early Mining in the Joplin District," *EMJ*, Vol. LVIII (December 1, 1894), 508.

46. Holibaugh, *Lead and Zinc Mining Industry*, 35.

47. "Lead Mining in Southwest Missouri," *Debow's Review*, Vol. XVIII (January, 1853), 389–91.

48. Swallow, *First and Second Annual Reports*, 163–64.

49. "Lead Mining in Southwest Missouri," *Debow's Review*, Vol. XVIII (January, 1853), 389–91.

50. Swallow, *First and Second Annual Reports*, 161.

51. That flatboating was widely used on this route for transporting a variety of products is evidenced by Muriel Wright in "Early Navigation and Commerce Along the Arkansas and Red Rivers in Oklahoma," *Chronicles of Oklahoma*, Vol. VIII (March, 1930), 63–68, and Daniel Dana, *Journal of a Tour in the Indian Territory, 1844*, 23.

52. "Mineral Wealth of Missouri," *Western Journal and Civilian*, Vol. VI (1853) 229–34. Also see Holibaugh, "Early Mining in the Joplin District," *EMJ*, Vol. LVIII (December 1, 1894), 508, and Winslow, *First and Second Annual Reports*, 230.

53. "Lead Mining in Southwest Missouri," *Debow's Review*, Vol. XVIII (January, 1853), 389–91.

54. Britton, *Pioneer Life*, 17–18.

55. Moseley, "Paper on the Lead Mines," *Western Journal and Civilian*, Vol. IV (1851), 412.

56. Britton, *Pioneer Life*, 18.

57. J. P. Benjamin, Acting Secretary of War to General Ben McCulloch, Richmond, Virginia, October 11, 1861, *War of Rebellion Records*, Series I, III, 717.

58. G. W. Clark to J. P. Benjamin, Acting Secretary of War, Fort Smith, Arkansas, October 14, 1861, *War of Rebellion Records*, Series I, III, 718.

59. Report of Lt. Col. Thomas Crittendon to Col. William Cloud, *War of Rebellion Records*, Series I, III, 718.

60. S. Waterhouse, *The Resources of Missouri*, 19.

III. DISTRICT DEVELOPMENT SINCE 1865

1. Tri-State Survey Committee, *A Preliminary Report on Living, Working, and Health Conditions in the Tri-State Mining Area, Missouri, Oklahoma, and Kansas*, 14.

2. Howard L. Conard, *Encyclopedia of the History of Missouri*, VI, 554.

3. Data for the roster of mining towns and their geographical location were extracted from Holibaugh, *The Lead and Zinc Mining Industry*, 28; North, *Jasper County*, 488–89; "Dayton Lead Mines," *EMJ*, Vol. XXII (July 22, 1876), 61; Livingston, *Jasper County*, I, 154;

Broadhead, *Report, 1873–1874*, 489; Haworth, Crane, and Rogers, *Special Report*, 19–20; Carthage *Weekly Banner*, March 27, 1882; *Historical Atlas*, 40; *Plat Book of Jasper County, Missouri*, 17; and *History of Newton, Lawrence, Barry, and McDonald Counties*, 28.

4. Data for the roster of mines were taken from "Chitwood Hollow," *EMJ*, Vol. LXIX (April 14, 1900), 535; D. Mathews, *Natural History of Jasper County*, 758; Wright, *Mining and Treatment*, 8–9; "Mining News," *EMJ*, Vol. CXVII (May 31, 1924), 894; and "Notes on Zinc-Lead Mining in Missouri," *EMJ*, Vol. XC (December 3, 1910), 1110.

5. Swallow, *First and Second Annual Reports*, 161.

6. William E. Connelley, *Quantrill and the Border Wars*, 421.

7. Nathan H. Parker, *Missouri As It Is in 1867—An Illustrated Historical Gazetteer of Missouri*, 51.

8. Nathan H. Parker, *The Missouri Handbook—Information for Capitalists and Immigrants*, 133.

9. Malcolm G. McGregor, *The Biographical Record of Jasper County, Missouri*, 32.

10. *Historical Atlas*, 19.

11. Joplin *Daily Herald*, May 8, 1880.

12. Winslow, *Lead and Zinc Deposits*, 294.

13. Livingston, *Jasper County*, I, 144–46.

14. Winslow, *Lead and Zinc Deposits*, 294.

15. Carthage *Weekly Banner*, August 10, 1871.

16. Livingston, *Jasper County*, I, 146–50.

17. Winslow, *Lead and Zinc Deposits*, 193.

18. Livingston, *Jasper County*, I, 151.

19. Joplin Chamber of Commerce, "Joplin, Missouri: History and Development" (Joplin, 1933), 2. Mimeographed.

20. Shaner, *The Story of Joplin*, 19.

21. North, *Jasper County*, 393.

22. *Ibid.*, 392–93.

23. Livingston, *Jasper County*, I, 152.

24. See Edwin C. McReynolds, "Joplin Lead Mining District," *Dictionary of American History* (ed. by James T. Adams), III, 181.

25. Shaner, *The Story of Joplin*, 24–25.

26. Joplin Chamber of Commerce, "Mining History of Joplin" (Joplin, 1950), 3. Mimeographed.

27. Livingston, *Jasper County*, I, 155.

28. North, *Jasper County*, 395.

29. Livingston, *Jasper County*, I, 154.

30. *Ibid.*, I, 154.

31. Eric Hedburg, "The Missouri and Arkansas Zinc Mines at the Close of 1900," *American Institute of Mining Engineering Transactions*, Vol. XXXI (1902), 397.

32. Livingston, *Jasper County*, I, 166.

33. Broadhead, *Report, 1873–1874*, 489.

34. North, *Jasper County*, 404.

35. Livingston, *Jasper County*, I, 175.

36. Joplin *Globe*, March 13, 1949.

37. Granby *Miner*, October 18, 1873.

38. Broadhead, *Report, 1873–1874*, 489. Today tourists traveling through Granby on U.S. Highway 60 are confronted in the town's center with a huge water tower emblazoned with the sign: "Welcome to Granby, First Mining Town in the Southwest."

39. Granby *Miner*, October 18, 1873.

40. Isaac Lippincott, "Industrial Influence of Lead in Missouri," *Journal of Political Economy*, Vol. XX (July, 1912), 713.

41. Livingston, *Jasper County*, I, 155.

42. Carthage *Weekly Banner*, March 27, 1884.

43. U.S. Bureau of the Census, *Tenth Census of the United States: 1880, Population*, I, 425.

44. North, *Jasper County*, 389.

45. "The town that Jack built" comes from the miner's name for zinc blende, "Jack" or "Blackjack," which along with lead furnished the basis for Joplin's existence in the first place.

46. Garland C. Broadhead, "Southwest Missouri Lead Interests," *EMJ*, Vol. XXV (February, 1883), 73.

47. Joplin remained famous as a liquor wholesale center for export through bootleggers into Kansas and Oklahoma before repeal of prohibition in those states.

48. *The Klondike of Missouri*, 13–14.

49. The *Annual Reports of the Commission of Indian Affairs* between 1880 and 1891 note the timber, grazing, and mineral resources, including lead, of the Quapaw Reservation, but no reference is made to actual mining before 1891. See *Annual Report of the Commissioner of Indian Affairs to the Secretary of the Interior for the Year 1891*.

50. G. W. Moseley, "Lead in Southwest Missouri," *Western Journal and Civilian*, Vol. XIII (January, 1855), 119.

51. Interview with H. A. Andrews, August 24, 1953, Miami, Oklahoma.

52. North, *Jasper County*, 253.

53. Broadhead, *Report, 1873–1874*, 441.

54. Luther C. Snider, *Preliminary Report on the Lead and Zinc of Oklahoma*, Oklahoma Geological Survey *Bulletin* No. 9, 61.

55. A. J. Martin, *Summarized Statistics of Production of Lead and Zinc in the Tri-State Mining District*, U.S. Bureau of Mines *Information Circular* No. 7383, 29.

56. *Daily Oklahoman*, August 22, 1897.

57. John S. Redfield, *Mineral Resources in Oklahoma*, Oklahoma Geological Survey *Bulletin* No. 42, 95.

58. Weidman, Williams, and Anderson, *Miami-Picher District*, 46.

59. Luther C. Snider, "Oklahoma Lead and Zinc Fields," *EMJ*, Vol.

XCII (December 23, 1911), 1228–30, and "Miami District of Oklahoma," *EMJ*, Vol. LXXXIX (January 8, 1910), 73.

60. Weidman, Williams, and Anderson, *Miami-Picher District*, 47.

61. Jesse A. Zook, "The Joplin District in 1914," *EMJ*, Vol. XCIX (January 9, 1915), 65.

62. "Editorial Correspondence," *EMJ*, Vol. CI (May 6, 1916), 834.

IV. TRI-STATE PROSPECTING

1. An early-day miner at Granby recounted in 1950 that "miners were always fooling round with fortune tellers. They'd draw a big ring on the map and the fortune teller would tell 'em whether or not there was lead inside the ring. Then they'd draw a ring inside o' that, and another one inside that one, and so on right down to a little bit of a ring. Then if a fortune teller would tell 'em there was lead inside that, they'd go out prospectin." Granby *News Herald*, Centennial Edition, June 1, 1950.

2. J. Frank Haley, "Churn Drilling in the Joplin District," *EMJ*, Vol. LXXXIX (June 4, 1910), 1150.

3. Bain, Van Hise, and Adams, *Preliminary Report*, 220–21.

4. W. R. Crane, "Methods of Prospecting in the Galena-Joplin District," *EMJ*, Vol. LXXII (September 21, 1903), 360.

5. Britton, *Pioneer Life*, 18–19.

6. Holibaugh, *The Lead and Zinc Mining Industry*, 6.

7. Haley, "Churn Drilling in the Joplin District," *EMJ*, Vol. LXXIX (June 4, 1910), 1150. Haworth found in the Galena, Kansas, camp that "early . . . scores of prospect shafts were sunk to a depth which in those days was considered the limit of profitable mining. After a lapse of fifteen years' idleness, a fortunate prospector began deepening one of these old shafts. He had gone less than four feet when he came upon one of the largest and richest bodies of ore ever yet discovered. In a few weeks' time the entire . . . valley was alive with prospectors, almost everyone of whom found large bodies of ore at about the same depth." Pumping lowered the water level and the production of this valley for the years 1896 and 1897 was worth $4,000,000. Haworth, *Annual Bulletin, 1897*, 29–30.

8. Wright, *Mining and Milling*, 5.

9. E. O. Bartlett, and P. L. Crossman, "Distribution of Lead and Zinc Ores Near Joplin, Missouri," *EMJ*, Vol. LXVII (March 18, 1899), 321.

10. F. H. Gartung, G. M. Burke, and O. W. Bilharz, "Prospecting, Development, and Mining," *American Mining Congress and American Institute of Mining and Metallurgical Engineers Convention Booklet* (September 28, 1931), 18. The cost of operating these early drills is estimated by Titcomb as follows: one drill man at $3.50 a day, $21.00 a week; one helper at $2.00 a day, $12.00 a week; water, wood, coal, and oil, $12.00 a week; depreciation on equipment, $10.00 a week; average

cost of drilling a foot, $1.25. From Harold A. Titcomb, "Cost of Prospecting by Drilling in Joplin District," *EMJ*, Vol. LXIX (April 7, 1900), 405.

11. Crane, "Methods of Prospecting and Mining," *EMJ*, Vol. LXXII (September 21, 1903), 360.

12. Titcomb, "Cost of Prospecting by Drilling," *EMJ*, Vol. LXIX (April 7, 1900), 405.

13. *Ibid.*, 405.

14. Crane, "Methods of Prospecting and Mining," *EMJ*, Vol. LXXII (September 21, 1903), 361.

15. Haworth, Crane, and Rogers, *Special Report*, 180.

16. W. F. Netzebrand, "An Example of Mining Lead and Zinc Ore at Picher, Oklahoma," *EMJ*, Vol. CXXVII (May 18, 1929), 793.

17. Haworth, Crane, and Rogers, *Special Report*, 181, and Crane, "Methods of Prospecting," *EMJ*, Vol. LXXII (September 21, 1903), 360.

18. Wright, *Mining and Milling*, 5.

19. Snider, *Preliminary Report*, 71–72.

20. J. R. Finlay, "Lead and Zinc Ores in Missouri," *EMJ*, Vol. LXXXVI (September 26, 1908), 605.

21. Finlay, *The Cost of Mining*, 314.

22. Titcomb, "Cost of Prospecting by Drill," *EMJ*, Vol. LXIX (April 7, 1900), 405.

23. Snider, *Preliminary Report*, 71–72, and W. F. Netzebrand, "An Example of Prospecting and Valuing a Lead-Zinc Deposit," *EMJ*, Vol. CXXVII (June 8, 1929), 915.

24. Wright, *Mining and Milling*, 5, and Finlay, "Lead and Zinc Ores," *EMJ*, Vol. LXXXVI (September 26, 1908), 608.

25. Weidman, Williams, and Anderson, *Miami-Picher District*, 94, Wright, *Mining and Milling*, 5–6, and Finlay, *Cost of Mining*, 94.

26. Frequently the drill struck water of such high acid content that, when pumped out, it was highly injurious to ordinary pipes. It was the practice to analyze the water and, when acid content registered high, the pumping apparatus was fitted with bronze fittings and a wooden discharge pipe. "Editorial Correspondence," *EMJ*, Vol. XCIX (February 7, 1915), 425.

27. Haley, "Churn Drilling in the Joplin District," *EMJ*, Vol. LXXXIX (June 4, 1910), 1150–51.

28. Frank W. Sansom, "Core Drills in the Joplin District," *EMJ*, Vol. XCI (April 1, 1911), 386.

29. "Editorial Correspondence," *EMJ*, Vol. XCIX (February 7, 1915), 425.

30. Netzebrand, "An Example of Mining Lead and Zinc," *EMJ*, Vol. CXXVII (May 18, 1929), 794. Also see Jackson et al., *Lead and Zinc Mining*, 53–54.

31. Weidman, Williams, and Anderson, *Miami-Picher District*, 94.

32. Netzebrand, "An Example of Mining Lead and Zinc," *EMJ*, Vol. CXXVII (May 18, 1929), 794.

33. Jackson et al., *Lead and Zinc Mining*, 53.

34. Haworth, Crane, and Rogers, *Special Report*, 113.

35. Crane, "Methods of Prospecting and Mining," *EMJ*, Vol. LXXII (September 21, 1903), 360, and Jackson et al., *Lead and Zinc Mining*, 48.

36. Netzebrand, "An Example of Mining Lead and Zinc," *EMJ*, Vol. CXXVII (May 18, 1929), 793.

37. "Mining News," *EMJ*, Vol. CXXII (October 2, 1926), 540, and "Editorial Correspondence," *EMJ*, Vol. XCVI (August 9, 1913), 278.

38. Snider, *Preliminary Report*, 15–16.

39. Geological Survey of Kansas, *Geophysical Investigations in the Tri-State Zinc and Lead Mining District*, 11–15.

40. Keener, *Method and Cost at the Hartley-Grantham Mine*, 2.

41. Jackson et al., *Lead and Zinc Mining*, 52.

42. Interview with Roy Moore, Purdy, Missouri, April 3, 1953.

43. Interview with Clarence Brown, Joplin, Missouri, April 2, 1953.

V. TRI-STATE MINING METHODS: THE EARLY PERIOD

1. Winslow, *Lead and Zinc Deposits*, 284.

2. Finlay, *The Cost of Mining*, 317.

3. Mine La Motte, worked for over two hundred years, has produced mineral worth $10,000,000. Southwestern mines, situated in more scattered deposits, have produced some rich strikes too, but nothing to match eastern Missouri lead and zinc statistics. The Hatton Mine on Center Creek Company land near Webb City, Missouri, produced over $300,000 worth of lead and zinc ores in eight years. The Victor Mining Company produced $275,000 worth of ores for the same period. The Paxton Mine of Joplin produced in six years $630,000 worth of ore. Winslow, *Lead and Zinc Deposits*, viii.

4. J. B. Guinn, "Missouri-Kansas Mining Methods and the Leasing System," *EMJ*, Vol. LXVIII (August 5, 1899), 154.

5. Missouri Bureau of Mines, *Fifteenth Annual Report of the State Lead and Zinc Mine Inspector, State of Missouri for 1901*, 84. The tradition of the Tri-State District being a "poor man's camp" possibly helps to explain the reluctance of local miners to affiliate with labor unions as late as 1935.

6. Granby *News Herald*, Centennial Edition, June 1, 1950.

7. "Mining News," *EMJ*, Vol. XLIX (February 8, 1890), 183.

8. Missouri Bureau of Mines, *Seventeenth Annual Report of the State Lead and Zinc Inspector, State of Missouri for 1903*, 29.

9. Harold A. Titcomb, "The Missouri-Kansas Zinc and Lead Mines," *EMJ*, Vol. LXX (July 28, 1900), 98.

10. Missouri Bureau of Mines, *Seventeenth Annual Report*, 116–17.

11. Moseley's mine on Shoal Creek reported in 1854 that three miners and a ground boss were employed to operate this famous early digging. Swallow, *First and Second Annual Reports,* 161.

12. Irene G. Stone, "The Lead and Zinc Field of Kansas," *Kansas Historical Collections,* Vol. VII (1902), 256.

13. Missouri Bureau of Mines, *Fifteenth Annual Report,* 24–25. At the prevailing local price of lead, 2½ cents a pound, the partners would each receive around $42 apiece for their labors. The world market price of lead in 1876 was 4.7 cents a pound. For a summary of lead prices a pound from 1771 to 1893, see Winslow, *Lead and Zinc Deposits,* 261.

14. Shaner, *Story of Joplin,* 8.

15. Granby *News Herald,* Centennial Edition, June 1, 1950.

16. Stone, "The Lead and Zinc Field of Kansas," *Kansas Historical Collections,* Vol. VII (1902), 254.

17. Shaner, *Story of Joplin,* 8.

18. Missouri Bureau of Mines, *Fifteenth Annual Report,* 25.

19. *Historical Atlas,* 19.

20. Doss Brittain, "Milling Sheet Ground Ore in the Joplin District," *EMJ,* Vol. LXXXIV (July 13, 1907), 61.

21. Britton, *Pioneer Life,* 20–21.

22. Interview with Ira Davis, April 1, 1953.

23. Shaner, *Story of Joplin,* 8.

24. Granby *News Herald,* Centennial Edition, June 1, 1950. Also see Brittain, "Ground Breaking in the Joplin District," *EMJ,* Vol. LXXXIV (August 10, 1907), 256.

25. Granby *News Herald,* Centennial Edition, June 1, 1950.

26. Broadhead, *Report, 1873–1874,* 490.

27. Edwin Higgins et al., *Siliceous Dust in Relation to Pulmonary Disease Among Miners in the Joplin District, Missouri,* U.S. Bureau of Mines *Bulletin* No. 132, 32.

28. Winslow, *Lead and Zinc Deposits,* 295.

29. Britton, *Pioneer Life,* 21–22.

30. Winslow, *Lead and Zinc Deposits,* 503.

31. Broadhead, *Report, 1873–1874,* 490.

32. Wright, *Mining and Milling,* 24.

33. Haworth, *Annual Bulletin, 1897,* 28.

34. Wright, *Mining and Milling,* 23–24.

35. Crane, "Methods of Prospecting and Mining," *EMJ,* Vol. LXXII (September 21, 1903), 361.

36. Haworth, *Annual Bulletin, 1897,* 28, and Broadhead, *Report, 1873–1874,* 491.

37. Joplin *Daily Herald,* May 4, 1880, and Granby *News Herald,* Centennial Edition, June 1, 1950.

38. Broadhead, *Report, 1873–1874,* 498.

39. Snider, "Oklahoma Lead and Zinc Fields," *EMJ,* Vol. XCII

(December 23, 1911), 1228, and Haworth, *Annual Bulletin, 1897*, 28.

40. Broadhead, *Report, 1873–1874*, 490.

41. Crane, "Methods of Prospecting and Mining," *EMJ*, Vol. LXXII (September 21, 1903), 361.

42. Nathaniel T. Allison, *History of Cherokee County, Kansas*, 128.

43. Crane, "Methods of Prospecting and Mining," *EMJ*, Vol. LXXII (September 21, 1903), 361.

44. Missouri Bureau of Mines, *Fifteenth Annual Report*, 83–84.

45. Granby *News Herald*, Centennial Edition, June 1, 1950.

46. Broadhead, *Report, 1873–1874*, 497.

47. Haworth, *Annual Bulletin, 1897*, 28.

48. Granby *News Herald*, Centennial Edition, June 1, 1950.

49. Wright, *Mining and Milling*, 24.

50. Swallow, *First and Second Annual Reports*, 161.

51. Broadhead, *Report, 1873–1874*, 491.

VI. TRI-STATE MINING METHODS: THE INTERMEDIATE PERIOD

1. Tyler, *From the Ground Up*, 24.

2. Oklahoma Bureau of Mines, *First Annual Report of the Chief Inspector of Mines, State of Oklahoma*, 17, and Holibaugh, *Lead and Zinc Mining Industry*, 28.

3. Bain, Van Hise, and Adams, *Preliminary Report*, 225.

4. Wright, *Mining and Milling*, 20.

5. D. F. Boardman, "Power Systems of Mines of the Joplin District," *EMJ*, Vol. LXXXVI (August 15, 1908), 327–29.

6. The Center Creek Company, operating in the Webb City field, had over two thousand feet of steam-line connections running from a central power plant to the pump shafts. "Mining News," *EMJ*, Vol. L (July 19, 1890), 253.

7. "Mining News," *EMJ*, Vol. L (August 30, 1890), 253.

8. Zook, "Conditions in the Joplin District," *EMJ*, Vol. LXXIX (January 12, 1905), 86.

9. "Electrical Power at Joplin," *EMJ*, Vol. LXXX (July 15, 1905), 64.

10. "Developments in the Joplin District," *EMJ*, Vol. XCVI (October 11, 1913), 684.

11. The cost of electrical power to the mines was 1¼ cents for one horsepower for each hour, plus a service charge of fifty dollars a month for each mine. This was cheaper than power could be generated by steam. Other advantages included the fact that less water was required, and the initial cost of electric motors was less than that of a steam plant. "Electrical Power at Joplin," *EMJ*, Vol. LXXX (July 15, 1904), 64.

12. Zook, "The Joplin District in 1905," *EMJ*, Vol. LXXXI (January 6, 1906), 14–15.

13. Wright made two studies of power use in the Tri-State District. His 1913 analysis disclosed that natural gas, electricity, coal, and oil were used as fuel for generating power, the operator seeking the cheapest power possible. "Of the mines under study, twenty-five used gas exclusively for fuel throughout the year, seven used electricity, coal was used at two. Gas was used in combination with other fuels at forty-five mines, electricity at thirteen, coal at sixteen, and oil at five. Electricity was used at all for lighting purposes. Coal and oil generally were used during the winter months when the gas pressure was lower. The mines using steam for power as a rule had two or three boilers of 100 to 150 horsepower capacity each. This plant supplied steam for compressors, hoisting engines, pumps and mills." Wright noted that several of the new mines were equipped entirely with electrical power. His 1918 study disclosed that natural gas was the most favored because of its cheapness, with coal and oil the least. See Wright, *Mining and Treatment*, 18, and Wright, *Mining and Milling*, 34.

14. Doss Brittain, "The Use of Natural Gas in the Joplin District," *EMJ*, Vol. LXXXVI (September 19, 1908), 568.

15. "Mining News," *EMJ*, Vol. CXXII (August 21, 1926), 308.

16. Stone, "The Lead and Zinc Field of Kansas," *Kansas Historical Collections*, Vol. VII (1902), 256.

17. North, *Jasper County*, 406, and Shaner, *The Story of Joplin*, 45.

18. Livingston, *Jasper County*, I, 201.

19. Allison, *History of Cherokee County, Kansas*, 135.

20. Missouri Bureau of Mines, *Fifteenth Annual Report*, 83–84.

21. "Mining News," *EMJ*, Vol. CXXII (August 21, 1926), 308.

22. H. W. Kitson, "The Mining Districts of Joplin and Southeast Missouri," *EMJ*, Vol. CV (February 23, 1918), 359.

23. "Mining News," *EMJ*, Vol. XLIX (April 12, 1890), 430.

24. Crane, "Recent Changes in Mining and Milling," *EMJ*, Vol. LXXIV (September 27, 1902), 405.

25. Wright, *Mining and Milling*, 23–24.

26. Brittain, "Ground Breaking in the Joplin District," *EMJ*, Vol. LXXXIV (August 10, 1907), 256.

27. Oklahoma Bureau of Mines, *First Annual Report*, 17.

28. Wright, *Mining and Milling*, 23–24.

29. "Mining News," *EMJ*, Vol. LXV (January 8, 1898), 51.

30. Wright, *Mining and Milling*, 25. Also see Titcomb, "The Missouri-Kansas Zinc and Lead Mines," *EMJ*, Vol. LXX (July 28, 1900), 99.

31. "Inclined Shafts at Joplin," *EMJ*, Vol. LXXXIX (January 22, 1910), 208.

32. Wright, *Mining and Treatment*, 19.

33. Haworth, Crane, and Rogers, *Special Report*, 373.

34. Missouri Bureau of Mines, *Fifteenth Annual Report*, 83–84.

35. Haworth, Crane, and Rogers, *Special Report*, 94.

36. Snider, *Preliminary Report*, 73. Also see G. W. Brigham, "Lead and Zinc Mining in Oklahoma in 1910," *EMJ*, Vol. XCI (January 7, 1911), 25, R. R. Heap, "A Geological Drainage Problem: Miami Camp," *EMJ*, Vol. XCVI (December 28, 1913), 1211, and Wright, *Mining and Milling*, 33.

37. Doss Brittain, "Pumping Problems of the Joplin District," *EMJ*, Vol. LXXXVI (August 1, 1908), 214–17.

38. Wright, *Mining and Milling*, 34.

39. Wright, *Mining and Treatment*, 19.

40. Beating the water has produced some Paul Bunyan-type stories among the miners. One recounts the exploits of fabulous "Hog Bed" Grigsby. At Granby, a large underground stream of water was struck in shafting and the pumps were not equal to the water flow. Mr. Grubb, a local miner and inventor, fashioned a special pump to bring the unusual flow under control. Hog Bed installed the impromptu pump in the shaft and operated it at the water level. On the up-stroke the water came up to Hog Bed's chin. He was reportedly six feet tall. Granby *News Herald*, Centennial Edition, June 1, 1950.

41. "Mining News," *EMJ*, Vol. LXX (December 15, 1900), 709.

42. Brittain, "Pumping Problems of the Joplin District," *EMJ*, Vol. LXXXVI (August 1, 1908), 214.

43. Wright, *Mining and Milling*, 34.

44. Snider, *Preliminary Report*, 16.

45. "Joplin-Miami District," *EMJ*, Vol. CIX (March 6, 1920), 629.

46. Brittain, "Ground Breaking in the Joplin District," *EMJ*, Vol. LXXXIV (August 10, 1907), 258, and Crane, "Methods of Prospecting and Mining," *EMJ*, Vol. LXXII (September 21, 1903), 361.

47. Wright, *Mining and Milling*, 26.

48. Wright, *Mining and Treatment*, 14–15.

49. Netzebrand, "An Example of Mining Lead and Zinc," *EMJ*, Vol. CXXVII (May 18, 1929), 795. Explosives came to the mines in fifty-pound wooden boxes, containing eighty cartridges each. The cost of a box of powder averaged $4.50. Dynamite was stored in surface or underground magazines. The operator generally kept fifteen to thirty boxes on hand at a time. The demand for powder in the district gave rise to three powder companies situated locally. Careful use of explosives was essential since, next to labor, explosives were the most important item in mining costs. See "Editorial Correspondence-Joplin," *EMJ*, Vol. CII (July 15, 1916), 157, and Wright, *Mining and Treatment*, 16.

50. Crane, "Methods of Prospecting and Mining," *EMJ*, Vol. LXXII (September 21, 1903), 362.

51. *Ibid.*

52. The Troup mine of the Carterville camp, one of the district's largest and heaviest producers, was operated at the two-hundred-foot level. Fifty-foot stopes were cut without adequate supports, creating

large, open chambers. Heavy rains soaked the surface, causing the roof to cave, taking the lives of three men. "Mining News," *EMJ*, Vol. LIII (June 4, 1892), 601.

53. H. K. Landis, "Zinc Mining in Missouri," *EMJ*, Vol. LXI (January 11, 1896), 39.

54. R. R. Hornor and H. E. Tufft, *Mine Timber, Its Selection, Storage, Treatment, and Use*, U.S. Bureau of Mines, 28.

55. "Mining News," *EMJ*, Vol. XC (September 10, 1910), 530.

56. Wright, *Mining and Treatment*, 18–19.

57. Snider, *Preliminary Report*, 16.

58. Edwin Higgins, "Shoveling and Tramming in the Joplin District," *EMJ*, Vol. C (October 23, 1915), 679. Also see Wright, *Mining and Treatment*, 17.

59. H. W. Kitson, "The Mining District of Joplin and Southeast Missouri," *EMJ*, Vol. CV (March 2, 1918), 411–16.

60. "Joplin Bucket Cars," *EMJ*, Vol. XCIII (May 18, 1912), 979–80.

61. Zook, "Zinc and Lead in the Joplin District in 1908," *EMJ*, Vol. LXXXVII (January 9, 1909), 71.

62. H. I. Young, "Mining Practice in the Joplin District," *EMJ*, Vol. CIV (October 6, 1907), 596.

63. Higgins, "Shoveling and Tramming in the Joplin District," *EMJ*, Vol. C (October 23, 1915), 689.

64. Joplin Chamber of Commerce, "Mining History of Joplin" (Joplin, 1950), 2. Mimeographed.

65. Wright, *Mining and Milling*, 24. Finlay found that the cost of shaft sinking to the 250-foot level was around four thousand dollars in 1909. See Finlay, *The Cost of Mining*, 315.

66. Claude T. Rice, "Joplin Hoisting Bucket Hooks," *EMJ*, Vol. XCIII (June 1, 1912), 1075.

67. "Routine of Joplin Hoisting," *EMJ*, Vol. XCIV (September 14, 1912), 491–92.

68. Joplin Chamber of Commerce, "Mining History of Joplin" (Joplin, 1950), 2. Mimeographed.

69. As late as 1912, horse-whim hoists were still widely used in smaller operations. See "Joplin Type of Horse Whim," *EMJ*, Vol. XCIV (August 3, 1912), 205.

70. Crane, "Recent Changes in Milling and Mining," *EMJ*, Vol. LXXIV (September 27, 1902), 405.

71. Brittain, "Ground Breaking in the Joplin District," *EMJ*, Vol. LXXXIV (August 10, 1907), 259.

72. "Mining News," *EMJ*, Vol. XC (July 30, 1910), 233.

VII. TRI-STATE MINING METHODS: THE MODERN PERIOD

1. Martin, *Statistics of Production of Lead and Zinc*, 4.

2. Weidman, Williams, and Anderson, *Miami-Picher District*, 95.

Wade Kurtz, a C.P.A. of Joplin, has furnished a sample of mine accounting in *ibid.*, 107.

Concentrate ton cost	*Zinc and lead operating cost*	*Percentage of total cost*
Labor, all classes	$11.0130	33.01
Salaries	1.6655	4.99
Explosives	1.7688	5.30
Supplies and repairs	4.0028	12.00
Power and power fuel	3.0340	9.10
Liability insurance	.5922	1.78
Fire and tornado insurance	.4583	1.37
Taxes (exclusive of income)	.2960	.89
Exploration and development	.0835	.25
Miscellaneous	1.8972	5.69
Total mining cost	$24.8113	74.38

	Zinc	*Lead*	*Lead and Zinc*	*Percentage*
Operating cost	$24.8113	$24.8113	$24.8113	74.38
Royalty	2.9774	6.0697	3.2394	9.71
Depreciation and depletion, or amortization of capital investment	5.3090	5.3090	5.3090	15.91
	$33.0977	$36.1900	$33.3597	100.00

The market price for lead concentrates in 1932 was $35.56, while zinc brought $18.25. The two metals are mined together, and according to column three, the combined cost of mining a ton of each was $33.35. A ton of each brought a total of $53.81 to the operator, thereby netting $20.46. Production statistics from Martin, *Summarized Statistics*, 21.

3. S. S. Clarke, "Mining Methods," *EMJ*, Vol. CXLIV (November, 1943), 89.

4. J. D. Forrester and F. A. Taylor, *A Comparative Analysis of Some Recent Mining Practices in the Tri-State Mining District*, Missouri School of Mines and Metallurgy *Technical Series Bulletin* No. 16, 19.

5. In spite of rapid advances, visiting mine authorities reacted with dismay at what they considered the Tri-State lag in mining method. One engineer from Butte, Montana, observed that "the Tri-State District uses a telephone pole for a head frame (hoist derrick) and a dinner pail for a cage (ore can)." S. S. Clarke, "Surface Plant at a Typical Operation of the Eagle-Picher Lead Company," *Mining Congress Journal*, Vol. XV (November, 1929), 865.

6. Lucien Eaton, "Seventy-five Years of Progress in Metal Mining," in *Seventy-Five Years of Progress in the Mineral Industry*, 77.

7. Mabel H. Draper, *Though Long the Trail*, 167.

8. H. S. Giessing, "Accident Prevention in the Tri-State District," *Mining Congress Journal*, Vol. XV (November, 1929), 854.

9. Clarke, "Mining Methods," *EMJ*, Vol. CXLIV (November, 1943), 81.

10. Forrester and Taylor, *Comparative Analysis*, 14–15.

11. O. N. Wampler, "Safety and Industrial Relations," *Mining Congress Journal*, Vol. XV (November, 1929), 877.

12. Oliver W. Keener, *Method and Cost at Barr Mine, Tri-State Zinc and Lead District*, U.S. Bureau of Mines *Information Circular* No. 6159, 7, and *Method and Cost at Hartley-Grantham Mine*, 6.

13. Jackson et al., *Lead and Zinc Mining and Milling*, 83.

14. Eaton, "Seventy-Five Years of Progress," in *Seventy-Five Years of Progress in the Mineral Industry*, 45.

15. Martin, *Statistics of Production of Lead and Zinc*, 11.

16. Forrester and Taylor, *Comparative Analysis*, 16–17.

17. Keener, *Method and Cost at Barr Mine*, 6.

18. Clarke, "Mining Methods," *EMJ*, Vol. CXLIV (November, 1943), 84–85.

19. Interview with Leo Nigh, Joplin, Missouri, December 28, 1950.

20. Clarke, "Mining Methods," *EMJ*, Vol. CXLIV (November, 1943), 85.

21. Joplin Chamber of Commerce, "Joplin, Missouri: History and Development" (Joplin, 1933), 3. Mimeographed.

22. Netzebrand, "An Example of Mining Lead and Zinc," *EMJ*, Vol. CXXVII (May 18, 1929), 796.

23. Clarke, "Mining Methods," *EMJ*, Vol. CXLIV (November, 1943), 88.

24. *Ibid.*, 86.

25. Forrester and Taylor, *Comparative Analysis*, 54.

26. *Joplin Globe*, August 1, 1948.

27. Clarke, "Mining Methods," *EMJ*, Vol. CXLIV (November, 1943), 89.

28. Field notes, Treece, Kansas, December 29, 1950.

29. Joplin *Globe*, March 20, 1949, and October 28, 1950.

30. Crane, "Recent Changes in Mining and Milling," *EMJ*, Vol. LXXIV (September 27, 1902), 405.

31. "Mining News," *EMJ*, Vol. XC (July 30, 1910), 233.

32. W. F. Netzebrand and Howard O. Gray, "An Open Pit Zinc-Lead Mine in the Tri-State District," *EMJ*, Vol. CXL (June, 1939), 52.

33. Joplin *Globe*, March 20, 1949.

34. *Ibid.*, October 28, 1950.

35. *Ibid.*, August 26, 1951.

36. Louis C. Brichta, *Exploration of O'Jack Mining Company Zinc and Lead Deposits*, U.S. Bureau of Mines *Report of Investigations* No. 3970, 3.

37. Louis C. Brichta, *Exploration for Zinc and Lead Ore, Phelps Lease,* U.S. Bureau of Mines *Report of Investigations* No. 3941, 7.
38. Ruhl, Allen, and Holt, *Zinc-Lead Ore Reserves of the Tri-State District,* 2.
39. Field notes, Galena, Kansas, December 30, 1950.
40. Field notes, Spurgeon, Missouri, August 25, 1951, and interviews with Roy Moore, Purdy, Missouri, April 4, 1953; Clarence Brown, Joplin, Missouri, April 2, 1953; and Ira Davis, Chitwood, Missouri, April 5, 1953.

VIII. TRI-STATE MILLING

1. Joplin Chamber of Commerce, "Mining History of Joplin" (Joplin, 1950), 2. Mimeographed.
2. Finlay, "Lead and Zinc Ores in Missouri," *EMJ,* Vol. LXXXVI (September 26, 1908), 608.
3. Granby *News Herald,* Centennial Edition, June 1, 1950.
4. Many of these heaps of discarded rock were subsequently treated in modern mills at a profit. Miners' children scrapped for mineral overlooked in these rock piles and earned from ten to thirty cents a day. See Livingston, *Jasper County,* I, 32, and II, 764. During the depression of the 1930's, some men made a meager living for their families by scrapping these rock piles for bits of lead and zinc ores.
5. Broadhead, *Report, 1873–1874,* 91.
6. *Ibid.,* 491–92.
7. Missouri Bureau of Mines, *Fifteenth Annual Report,* 24–25.
8. Britton, *Pioneer Life,* 20–21.
9. Holibaugh, *Lead and Zinc Mining Industry,* 8.
10. Winslow, *Lead and Zinc Deposits,* 284.
11. Snider, *Preliminary Report,* 20.
12. Haworth, *Annual Bulletin for 1897,* 28.
13. Joplin Chamber of Commerce, "Mining History of Joplin" (Joplin, 1950), 3. Mimeographed.
14. Stone, "The Lead and Zinc Field of Kansas," *Kansas Historical Collections,* Vol. VII (1902), 254.
15. Shaner, *The Story of Joplin,* 9.
16. Titcomb, "The Missouri-Kansas Zinc and Lead Mines," *EMJ,* Vol. LXX (July 28, 1900), 98.
17. *The Klondike of Missouri,* 14.
18. F. L. Clerc, "Lewis Bartlett Process at Lone Elm," *EMJ,* Vol. XL (July 4, 1885), 4.
19. Broadhead, *Report, 1873–1874,* 499.
20. Stone, "The Lead and Zinc Field of Kansas," *Kansas Historical Collections,* Vol. VII (1902), 254.
21. "Mining News," *EMJ,* Vol. XLIX (March 29, 1890), 367.
22. Allison, *History of Cherokee County,* 128.
23. Stone, "The Lead and Zinc Field of Kansas," *Kansas Historical Collections,* Vol. VII (1902), 255.

24. Allison, *History of Cherokee County*, 128.
25. Broadhead, *Report, 1873–1874*, 499.
26. Stone, "The Lead and Zinc Field of Kansas," *Kansas Historical Collections*, Vol. VII (1902), 256.
27. Granby *News Herald*, Centennial Edition, June 1, 1950.
28. North, *Jasper County*, 474.
29. "Mining News," *EMJ*, Vol. L (August 23, 1890), 227.
30. Missouri Bureau of Mines, *Fifteenth Annual Report*, 25.
31. Finlay, *The Cost of Mining*, 316.
32. Bain, Van Hise, and Adams, *Preliminary Report*, 226. The concentrates were expressed in terms of zinc recovery since, by 1900, the ratio of lead to zinc mined and milled was about one to five. Because of the irregularity of the deposits, a single mine might in one drift produce 30 per cent ore and in another drift produce 4 per cent or less. The most consistent producer in the field at this time was the John Jackson mine near Chitwood which ran 22 per cent ore throughout.
33. Finlay, "Lead and Zinc Ores in Missouri," *EMJ*, Vol. LXXXVI (September 26, 1908), 608.
34. "Joplin Mill Practice," *EMJ*, Vol. LXXVIII (October 13, 1904), 579–80.
35. Bain, Van Hise, and Adams, *Preliminary Report*, 227.
36. "Miami District, Oklahoma," *EMJ*, Vol. LXXXIX (January 8, 1910), 73. A Joplin Mill of ten tons an hour could be built for $6,000 to $7,000 in 1904. An eight-ton mill in southeast Missouri cost $30,000. The milling cost a ton was about the same. The Joplin Mill was temporary while the southeastern mill was built to last.
37. Brittain, "Milling Sheet Ground Ore," *EMJ*, Vol. LXXXIV (July 13, 1907), 63.
38. H. W. Kitson, "The Mining District of Joplin and Southeast Missouri," *EMJ*, Vol. CV (April 20, 1918), 727.
39. Claude T. Rice, "A Joplin Car for Boulders," *EMJ*, Vol. XCIII (April 6, 1912), 690.
40. John S. Redfield, *Mineral Resources in Oklahoma*, Oklahoma Geological Survey *Bulletin* No. 42, 93.
41. Snider, *Preliminary Report*, 17.
42. Finlay, *Cost of Mining*, 317.
43. Wright, *Mining and Treatment*, 26.
44. George T. Cooley, "Dressing Zinc and Lead Ores in Southwest Missouri and Southeast Kansas," *EMJ*, Vol. LVIII (July 7, 1894), 9.
45. Snider, *Preliminary Report*, 18.
46. Wright, *Mining and Treatment*, 27.
47. W. R. Crane, "Slime Treatment in the Joplin Region," *EMJ*, Vol. LXXVII (April 28, 1904), 683.
48. Snider, *Preliminary Report*, 18.
49. "Joplin Type of Bucket Elevators," *EMJ*, Vol. XCIV (November 23, 1912), 977.

50. Joplin *Globe*, June 21, 1953.
51. Brittain, "Milling Sheet Ground Ore," *EMJ*, Vol. LXXXIV (July 13, 1907), 61.
52. Missouri Bureau of Mines, *Seventeenth Annual Report of the State Lead and Zinc Mine Inspector, State of Missouri, 1903*, 110–11.
53. Zook, "The Joplin District in 1905," *EMJ*, Vol. LXXXI (January 6, 1906), 15.
54. Lucius Wittich, "Flotation in the Joplin District," *EMJ*, Vol. CII (July 1, 1916), 45–46.
55. C. O. Anderson and Elmer Isern, "Milling," *American Mining Congress and American Institute of Mining and Metallurgical Engineers Convention Booklet* (September 28, 1931), 23.
56. F. W. Sansom, "Milling Practice at the Eagle Picher Lead Company," *Mining Congress Journal*, Vol. XV (November, 1929), 869.
57. Joplin Chamber of Commerce, "Mining History of Joplin" (Joplin, 1950), 2. Mimeographed.
58. C. O. Anderson, "Growth of Flotation in the Tri-State District," *EMJ*, Vol. XCI (June 12, 1926), 982.
59. Joplin Chamber of Commerce, "Mining History of Joplin" (Joplin, 1950), 3. Mimeographed.
60. Snider, *Preliminary Report*, 18.
61. Wittich, "Flotation in the Joplin District," *EMJ*, Vol. CII (July 1, 1916), 45–46.
62. "News from Washington," *EMJ*, Vol. CXXII (October 9, 1926), 584.
63. H. D. Keiser, "Retreating Coarse Tailings in the Tri-State District," *EMJ*, Vol. CXXIII (April 9, 1927), 596.
64. W. F. Netzebrand, "Surface Transport of Ore and Tailings in the Tri-State District," *EMJ*, Vol. CXXXVI (December, 1935), 601.
65. W. F. Netzebrand, "Truck Transport Serves Tri-State Mining," *EMJ*, Vol. CXLI (January, 1940), 34.
66. Evan Just, "Tri-State Strives to Maintain Output," *EMJ*, Vol. CXLIII (December, 1942), 69.
67. Wright, *Mining and Milling*, 11.
68. Wright, *Mining and Treatment*, 32.
69. "Miami District, Oklahoma," *EMJ*, Vol. LXXXIX (January 8, 1910), 73.
70. Snider, "Oklahoma Lead and Zinc Fields," *EMJ*, Vol. XCII (December 23, 1911), 1229.
71. A. L. Morford Journal entries for September, 1911, Galena, Kansas, 16.
72. Elmer Isern, "Central Milling in the Tri-State District," *EMJ*, Vol. CXXXI (January 26, 1931), 49.
73. "History of Eagle Picher," *EMJ*, Vol. CXLIV (November, 1943), 72.
74. H. A. Gray and R. J. Stroup, "Transportation," *EMJ*, Vol. CXLIV (November, 1943), 90–92.

IX. TRI-STATE SMELTING

1. See Witcombe, *All About Mining*, 151–57, and A. D. Terrell, "Smelting Practices," *American Mining Congress and American Institute of Mining and Metallurgical Engineers Convention Booklet* (September 28, 1931), 28–32, for the metallurgy of lead and zinc.

2. Doss Brittain, "History of Smelting in the Joplin District," *EMJ*, Vol. LXXXIV (November 9, 1907), 861.

3. See Holibaugh, *Lead and Zinc Mining Industry*, 32, for a description of pioneer furnaces.

4. Brittain, "History of Smelting," *EMJ*, Vol. LXXXIV (November 9, 1907), 861.

5. *Historical Atlas*, 15.

6. This step in smelting was later designated as roasting.

7. Schoolcraft, *A View*, 93.

8. See Winslow, *Lead and Zinc Deposits*, 206.

9. Brittain, "History of Smelting," *EMJ*, Vol. LXXXIV (November 9, 1907), 861.

10. Winslow, *Lead and Zinc Deposits*, 199.

11. "Lead Mining in Southwest Missouri," *DeBow's Review*, Vol. XVIII (January, 1853), 389–91.

12. William S. Moseley, "Smelting Lead in Southwest Missouri," *Western Journal and Civilian*, Vol. III (September, 1850), 412–13.

13. Swallow, *First and Second Annual Reports*, 163.

14. William H. Pulsifer, *Notes for a History of Lead and an Inquiry into the Development of the Manufacture of White Lead and Lead Oxides*, 1150–51.

15. Winslow, *Lead and Zinc Deposits*, 201.

16. Swallow, *First and Second Reports*, 163.

17. "Lead Mining in Southwest Missouri," *Debow's Review*, Vol. XVIII (January, 1853), 389–91.

18. Winslow, *Lead and Zinc Deposits*, 206.

19. "Lead Mining in Southwest Missouri," *Debow's Review*, Vol. XVIII (January, 1853), 389–91.

20. Livingston, *Jasper County*, I, 31.

21. "Lead Mining in Southwest Missouri," *Debow's Review*, Vol. XVIII (January, 1853), 389–91.

22. "The blast for Harklerode's furnaces is produced by waterpower derived from a large spring near the furnace. The charcoal used for smelting the lead costs about $.033 per bushel; according to the following estimate of expense for making 1600 bushels (25 cords of wood at $.75 per cord costs $18.75). One hand one month at $25.00 equals $25.00 and a team for hauling the wood $10.00 totaling $53.75. Ten bushels make a ton of lead. The mineral costs about $20.00 per thousand." Swallow, *First and Second Reports*, 163.

23. Through 1854, most of the lead produced in the district was sent to New York and Boston via Spring River flatboat to the Grand River

and the Arkansas, through the Indian Country to Fort Smith, then by steamboat to New Orleans. Moseley, "Smelting Lead in Southwest Missouri," *Western Journal and Civilian*, Vol. III (September, 1950), 412–13.

24. *Historical Atlas*, 22.

25. Brittain, "History of Smelting," *EMJ*, Vol. LXXXIV (November 9, 1907), 863.

26. See E. R. Buckley and H. A. Buehler, *The Geology of the Granby Area*, 13, for a description of early smelting at Granby.

27. "History of Eagle Picher," *EMJ*, Vol. CXLIV (November, 1943), 70.

28. Broadhead, *Report, 1873–1874*, 489.

29. Swallow, *First and Second Reports*, 163–64.

30. North, *Jasper County*, 392.

31. "History of Eagle Picher," *EMJ*, Vol. CXLIV (November, 1943), 70, and Brittain, "History of Smelting," *EMJ*, Vol. LXXXIV (November 9, 1907), 862.

32. Winslow, *Lead and Zinc Deposits*, 206. Each furnace was operated by two smeltermen, who put in a twelve-hour shift in winter, but because of the intense heat of the furnace, coupled with higher temperatures, worked only eight hours in the summer. See Charles P. Williams, *Industrial Report on Lead, Zinc, and Iron*, 37.

33. Scotch Eye was a Wisconsin and Missouri smelting term for furnace. Brittain, "History of Smelting," *EMJ*, Vol. LXXXIV (November 9, 1907), 861.

34. Clerc, "Lewis Bartlett Process at Lone Elm," *EMJ*, Vol. XL (July 4, 1885), 4.

35. Winslow, *Lead and Zinc Deposits*, 206.

36. Buckley and Buehler, *Geology of Granby Area*, 113.

37. Winslow, *Lead and Zinc Deposits*, 207.

38. Moses Austin to the President of the United States Concerning Lead Mines in Upper Louisiana, Mine a Burton, February 13, 1804, *American State Papers, Public Lands*, I, 188.

39. Broadhead, *Report, 1873–1874*, 497.

40. Winslow, *Lead and Zinc Deposits*, 207.

41. Brittain, "History of Smelting," *EMJ*, Vol. LXXXIV (November 9, 1907), 862.

42. Haworth, *Annual Bulletin, 1897*, 32.

43. Brittain, "History of Smelting," *EMJ*, Vol. LXXXIV (November 9, 1907), 861, 863.

44. Livingston, *Jasper County*, II, 592.

45. Clerc, "Lewis Bartlett Process at Lone Elm," *EMJ*, Vol. XL (July 4, 1885), 4.

46. North, *Jasper County*, 468.

47. Holibaugh, *Lead and Zinc Mining Industry*, 40.

48. "The Joplin White Lead Works," *EMJ*, Vol. XXV (June 8, 1878),

394. In 1878, Bartlett and Lewis sold the exclusive right to process lead fume into white lead paint in Jasper, Newton, and Cherokee counties to the Lone Elm Smelting Company and then went on to Colorado to establish a white-lead plant there. Brittain, "History of Smelting," *EMJ*, Vol. LXXXIV (November 9, 1907), 864.

49. Winslow, *Lead and Zinc Deposits*, 210.

50. Livingston, *Jasper County*, I, 169.

51. Holibaugh, "Early Mining in the Joplin District," *EMJ*, Vol. LVIII (December 1, 1898), 508.

52. "Mining News," *EMJ*, Vol. XXXIII (January 28, 1882), 56.

53. Brittain, "History of Smelting," *EMJ*, Vol. LXXXIV (November 9, 1907), 864.

54. Broadhead, *Report, 1873–1874*, 492.

55. "Mining News," *EMJ*, Vol. XXVI (July 6, 1878), 9.

56. Winslow, *Lead and Zinc Deposits*, 210.

57. "Mine Owners of Joplin," *EMJ*, Vol. L (July 26, 1890), 96.

58. Swallow, *First and Second Reports*, 161.

59. Granby *News Herald*, Centennial Edition, June 1, 1950.

60. Walter Williams, *The State of Missouri*, 167–68.

61. Otto Ruhl and O. W. Bilharz, "History of the Tri-State District," *American Mining Congress and American Institute of Mining and Metallurgical Engineers Convention Booklet* (September 28, 1931), 6.

62. "History of Eagle Picher," *EMJ*, Vol. CXLIV (November, 1943), 70.

63. Granby *Miner*, March 28, 1874.

64. Haworth, *Annual Bulletin, 1897*, 33.

65. Granby *News Herald*, Centennial Edition, June 1, 1950.

66. Haworth, *Annual Bulletin, 1897*, 33.

67. Winslow, *Lead and Zinc Deposits*, 234.

68. Terrell, "Smelting Practices," *American Mining Congress and American Institute of Mining and Metallurgical Engineers Convention Booklet* (September 28, 1931), 30.

69. North, *Jasper County*, 472.

70. Winslow, *Lead and Zinc Deposits*, 234.

71. Haworth, *Annual Bulletin, 1897*, 33–34.

72. Terrell, "Smelting Practices," *American Mining Congress and American Institute of Mining and Metallurgical Engineers Convention Booklet* (September 28, 1931), 29.

73. "Mining News," *EMJ*, Vol. XXXIV (October 7, 1882), 191.

74. John R. Holibaugh, "Zinc and Lead Mining in Missouri and Kansas in 1895," *EMJ*, Vol. LX (July 27, 1895), 79.

75. Shaner, *The Story of Joplin*, 77.

76. Holibaugh, "Lead and Zinc Mining Industry of Missouri and Kansas in 1892," *EMJ*, Vol. LV (February 18, 1893), 151.

77. "History of Eagle Picher," *EMJ*, Vol. CXLIV (November, 1943), 70.

78. "Editorial Correspondence," *EMJ*, Vol. XCIX (April 24, 1915), 756.

79. "Zinc Smelting in Kansas," *EMJ*, Vol. XC (October 15, 1910), 748.

80. "Zinc Smelting in Kansas and Oklahoma," *EMJ*, Vol. XCIII (April 27, 1912), 850.

81. Walter R. Ingalls, "Spelter Statistics for 1914," *EMJ*, Vol. XCIX (March 6, 1915), 453.

82. Joplin Chamber of Commerce, "Mining History of Joplin" (Joplin, 1950), 3. Mimeographed. Also see Joplin *Globe*, August 1, 1948.

X. TRI-STATE LAND TENURE AND USE SYSTEMS

1. See Chapter I for district geology. Especially note Finlay, *The Cost of Mining*, 317, and Bain, Van Hise, and Adams, *Preliminary Report*, 227.

2. Britton, *Pioneer Life*, 21.

3. Broadhead, *Report, 1873–1874*, 490.

4. *Missouri Mining Code, Revised Statutes of Missouri, 1899*, II, 2039–45.

5. In 1950 a limited amount of small-tract leasing was still practiced in the district. One change was that a suspension of active mining must not exceed twenty-one days. Interview with Roy Moore, Purdy, Missouri, April 4, 1953.

6. See Broadhead, *Report, 1873–1874*, 490, for a more detailed discussion of early leasing practices.

7. Carthage *Weekly Banner*, March 7, 1872.

8. *Kansas Mining Code, General Statutes of Kansas Annotated, 1949*, 1442–52, and *Oklahoma Lead and Zinc Mining Code, Oklahoma Statutes Annotated*, 622–36.

9. Snider, *Preliminary Report*, 15.

10. Stone, "The Lead and Zinc Field of Kansas," *Kansas Historical Collections*, Vol. VII (1902), 255–56.

11. Granby *News Herald*, Centennial Edition, June 1, 1950.

12. Finlay, *The Cost of Mining*, 317.

13. Joplin *Daily Herald*, May 4, 1880.

14. See Holibaugh, *The Lead and Zinc Mining Industry*, for a discussion of land division.

15. Witcombe, *All About Mining*, 157.

16. "New Mineral Discoveries," *Western Quarterly*, Vol. I (July, 1868), 89–92.

17. Broadhead, *Report, 1873–1874*, 496.

18. Missouri Bureau of Mines, *Fifteenth Annual Report*, 25.

19. "Mining News," *EMJ*, Vol. XLIX (February 8, 1890), 183.

20. Livingston, *Jasper County*, 178.

21. "Mining News," *EMJ*, Vol. XLVI (August 25, 1888), 425.

22. Broadhead, *Report, 1873–1874*, 448–57.
23. North, *Jasper County*, 490–91.
24. "The Joplin Zinc and Lead District in 1901," *EMJ*, Vol. LXXIII (January 14, 1902), 29.
25. "Mining News," *EMJ*, Vol. XLIX (March 1, 1890), 257.
26. "Mining News," *EMJ*, Vol. LXI, (April 4, 1896), 333.
27. "Mining News," *EMJ*, Vol. XLIX (March 1, 1890), 257.
28. "Mining News," *EMJ*, Vol. LIII (June 25, 1892), 672.
29. "Mining News," *EMJ*, Vol. LVII (January 27, 1892), 87.
30. Missouri Bureau of Mines, *Fifteenth Annual Report*, 11–13.
31. See Missouri Bureau of Mines, *Fifteenth Annual Report* and *Seventeenth Annual Report* for a more detailed listing of the land and royalty companies operating in the district.
32. Holibaugh, *The Lead and Zinc Mining Industry*, 13.
33. Interview with Leroy Adair, Granby, Missouri, August 13, 1951.
34. Granby *News Herald*, Centennial Edition, June 1, 1950.
35. "The Missouri Lead and Zinc Company's Plant," *EMJ*, Vol. LXIX (June 2, 1900), 649.
36. Picher *Tri-State Tribune*, December 21, 1951.
37. Martin, *Statistics of Production of Lead and Zinc*, 9.
38. "Miami District of Oklahoma," *EMJ*, Vol. LXXXIX (January 8, 1910), 73–74.
39. Wright, *Mining and Treatment*, 41.
40. Wright, *Mining and Milling*, 15.
41. Wright, *Mining and Treatment*, 41.
42. Lane Carter, "Economic Conditions in the Joplin District," *EMJ*, Vol. XC (October 15, 1910), 760.
43. Guinn, "Missouri-Kansas Mining Methods," *EMJ*, Vol. LXVIII (August 5, 1899), 154.
44. Nathan H. Parker, *Missouri as It Is in 1867—An Illustrated Historical Gazetteer of Missouri*, 285, and *History of Newton, Lawrence, Barry, and McDonald Counties of Missouri*, 200.
45. Winslow, *Lead and Zinc Deposits*, vii–ix.
46. "Mining News," *EMJ*, Vol. CXXI (March 6, 1926), 415.
47. *United States Statutes at Large*, XXX (Washington, 1899), 72.
48. C. F. Williams, "The Leasing System," *American Mining Congress and American Institute of Mining and Metallurgical Engineers Convention Booklet* (September 28, 1931), 10.
49. Weidman, Williams, and Anderson, *Miami-Picher Zinc-Lead District*, 92.
50. "Mining News," *EMJ*, Vol. CXXI (March 6, 1926), 415.
51. Interview with H. A. Andrews, Miami, Oklahoma, August 24, 1953.
52. "Mining News," *EMJ*, Vol. CXVI (November 3, 1923), 780.
53. Interview with H. A. Andrews, Miami, Oklahoma, August 24,

1953. Also see U.S. Bureau of Indian Affairs, *Lead and Zinc Mining Operations and Leases, Quapaw Agency: Regulations of the Indian Service.*

54. Weidman, Williams, and Anderson, *Miami-Picher District,* 92.

55. C. F. Williams, "The Tri-State District from the Air," *Mining Congress Journal,* Vol. XV (November, 1929), 846.

56. W. R. Crane and G. I. Adams, "Lead and Zinc Mining in the Quapaw District," *Mines and Minerals,* Vol. X (May, 1907), 446.

57. A. L. Morford Journal entries for December, 1907, Galena, Kansas, 41.

58. "Opposition to Removing Quapaw Restrictions Develops," *EMJ,* Vol. CIX (June 19, 1920), 1377.

59. *United States Statutes at Large,* XLI, 1225–48.

60. "News from Washington," *EMJ,* CXIII (March 18, 1922), 463.

61. "Mining News," *EMJ,* Vol. CXIII (June 3, 1922), 980.

62. Gertrude C. Bonnin et al., *Oklahoma's Poor Rich Indians.*

63. "Mining News," *EMJ,* Vol. CXXII (October 9, 1926), 583.

64. *Whitebird et al.* v. *Eagle Picher Lead Company, Federal Reporter*: Second Series, XXVIII (St. Paul, 1929), 200–205. For editorial comment see "Mining News," *EMJ,* Vol. CXXVI (September 22, 1928), 467, and Blackwell *Morning Tribune,* September 11, 1928.

65. Interview with H. A. Andrews, Miami, Oklahoma, August 24, 1953.

66. See Chapter IX for a more detailed account of central milling. For further details on the problems inherent in royalty leases, and their relation to mining and milling, see Morford, "Zinc Mining at Galena, Kansas," *EMJ,* Vol. XCII (December 16, 1911), 1171; Kitson, "The Mining Districts of Joplin and Southeast Missouri," *EMJ,* Vol. CIV (December 22, 1917), 360; Elmer Isern, "Central Milling," *EMJ,* Vol. CXLIV (November, 1943), 93–105; Isern, "Central Milling in the Tri-State District," *EMJ,* Vol. CXXXI (January 26, 1931), 49–54; Phil R. Coldren, "Joplin-Miami District." *EMJ,* Vol. CXIII (February 18, 1922), 307; Charles W. Burgess, "Mining Costs in the Missouri-Kansas District," *Mining and Engineering World,* Vol. XXXVIII (April, 1913), 801–805; Titcomb, "The Missouri-Kansas Zinc and Lead Mines," *EMJ,* Vol. LXX (July 28, 1900), 98.

XI. TRI-STATE MINE FINANCE

1. Walter R. Ingalls, *World Survey of the Zinc Industry,* 70.

2. Granby *News Herald,* Centennial Edition, June 1, 1950.

3. Livingston, *Jasper County,* II, 655.

4. Missouri Bureau of Labor Statistics, *Eleventh Annual Report of the Missouri Bureau of Labor Statistics,* 336.

5. "Mining News," *EMJ,* Vol. L (July 19, 1890), 82.

6. Witcombe, *All About Mining,* 157.

7. Bain, Van Hise, and Adams, *Preliminary Report*, 227.

8. Wright, *Mining and Treatment*, 26, and *Mining and Milling*, 13–14.

9. Frank Nicholson, "The Missouri-Kansas Zinc District," *EMJ*, Vol. LXVII (June 10, 1899), 680.

10. Carter, "Economic Conditions in the Joplin District," *EMJ*, Vol. XC (October 15, 1910), 759.

11. Ingalls, *World Survey*, 70.

12. "Development of the Joplin District," *EMJ*, Vol. LXVI (June 18, 1898), 725.

13. "Mining News," *EMJ*, Vol. L (August 30, 1890), 253.

14. "Mining News," *EMJ*, Vol. XLIX (March 1, 1890), 257.

15. "Mining News," *EMJ*, Vol. XLIX (May 10, 1890), 545.

16. "Mining News," *EMJ*, Vol. LIII (June 18, 1892), 648. Also see "Mining News," *EMJ*, Vol. LIII (June 25, 1892), 672.

17. "Mining News," *EMJ*, Vol. LXXXIX (October 19, 1907), 756.

18. "Mining News," *EMJ*, Vol. LIII (June 25, 1892), 672.

19. "Mining News," *EMJ*, Vol. XLIX (March 15, 1890), 318.

20. "Mining News," *EMJ*, Vol. LVII (January 27, 1894), 87.

21. Granby *News Herald*, Centennial Edition, June 1, 1950.

22. Granby *Miner*, October 18, 1873.

23. Missouri Bureau of Mines, *Seventeenth Annual Report*, 28.

24. "History of Eagle Picher," *EMJ*, Vol. CXLIV (November, 1943), 70.

25. North, *Jasper County*, 488–89.

26. Swallow, *The First and Second Annual Reports*, 161.

27. Holibaugh, *The Lead and Zinc Mining Industry*, 35.

28. Cumulative data on mining enterprises were compiled from "Mining News," *EMJ*, Vol. XLVI (August 25, 1888), 425; Holibaugh, *The Lead and Zinc Industry*, 13; North, *Jasper County*, 491; Livingston, *Jasper County*, I, 150; "Center Creek Mining Company," *EMJ*, Vol. XLVII (March 30, 1889), 30; Missouri Bureau of Mines, *Seventeenth Annual Report*, 38; Missouri Bureau of Mines, *Fifteenth Annual Report*, 84; and Snider, *Preliminary Report*, 28.

29. "Mining News," *EMJ*, Vol. LXIX (June 9, 1900), 689, and "Mining News," *EMJ*, Vol. LXIX (June 16, 1900), 719.

30. Missouri Bureau of Mines, *Fifteenth Annual Report*, 86.

31. *Ibid.*, 26.

32. Livingston, *Jasper County*, II, 571. Also see William R. and Mabel Draper, *Old Grubstake Days in Joplin*, 13.

33. "By the Way," *EMJ*, Vol. XCII (December 9, 1911), 1116.

34. "Mining News," *EMJ*, Vol. XLVII (April 20, 1889), 375.

35. Missouri Bureau of Mines, *Fifteenth Annual Report*, 77.

36. "Mining News," *EMJ*, Vol. LXI (February 1, 1896), 117.

37. Otto Ruhl, *The Rationale of Investment in Zinc Mining*, 18.

38. Martin, *Statistics of Production of Lead and Zinc*, 16.

39. Haworth, Crane, and Rogers, *Special Report*, 31.
40. "Mining News," *EMJ*, Vol. XC (December 3, 1910), 1110.
41. Wright, *Mining and Treatment*, 6, and *Mining and Milling*, 14.
42. Otto Ruhl, "Keeping Mining Costs at Joplin," *EMJ*, Vol. XCII (December 9, 1911), 1133.
43. Ingalls, *World Survey*, 67.
44. "Editorial Correspondence," *EMJ*, Vol. CV (April 20, 1918), 768.
45. Foster S. Naething, "The Oklahoma-Kansas-Missouri Zinc-Lead Field," *EMJ*, Vol. CXXII (October 16, 1926), 607.
46. Martin, *Statistics of Production of Lead and Zinc*, 22–28.
47. "History of Eagle Picher," *EMJ*, Vol. CXLIV (November, 1943), 68.
48. *Ibid.*, 70.
49. Edward D. Morton, "Arizona Operation," *EMJ*, Vol. CXLIV (November, 1943), 107.
50. D. C. MacKallor, "Mexican Operation," *EMJ*, Vol. CXLIV (November, 1943), 106.
51. Shaner, *The Story of Joplin*, 120.
52. "History of Eagle Picher," *EMJ*, Vol. CXLIV (November, 1943), 70.
53. Miami *Daily Record*, October 23, 1938.
54. *Ibid.*
55. Joplin *Globe*, September 29, 1951.
56. Tri-State Ore Producers' Association, "Companies Operating in the Tri-State District of Oklahoma, Kansas, and Missouri" (Picher, 1946), 1–2. Mimeographed.
57. U.S. Congress, House, Hearings Before the Select Committee on Small Business, *Problems in the Metal-Mining Industry (Lead, Zinc, and Other Metals)*, 83 Cong., 1 sess., 1953, 515.
58. Tyler, *From the Ground Up*, 148.
59. L. H. McColgin, "Marketing the Concentrates," *American Mining Congress and American Institute of Mining and Metallurgical Engineers Convention Booklet* (September 28, 1931), 27.
60. Wright, *Mining and Milling*, 120.
61. Evans W. Buskett, "Sampling and Buying Ore in the Joplin District," *EMJ*, Vol. LXXXVI (July 25, 1908), 190.
62. W. George Waring, "The Purchase of Zinc Ores in the Joplin District," *EMJ*, Vol. LXX (July 14, 1900), 38–39.
63. Buskett, "Sampling and Buying Ore," *EMJ*, Vol. LXXXVI (July 25, 1908), 190.
64. Missouri Bureau of Mines, *Seventeenth Annual Report*, 11.
65. Miami *Daily Record*, April 26, 1953.
66. "Mining News," *EMJ*, Vol. L (November 8, 1890), 553.
67. "The Ore Market of the Joplin District in 1896," *EMJ*, Vol. LXIII (January 2, 1897), 20.

68. "Mining News," *EMJ*, Vol. LXXI (March 16, 1901), 345.
69. "Mining News," *EMJ*, Vol. XLIX (April 19, 1890), 454.
70. A. Heckscher, "Zinc and Zinc Oxide in 1898," *EMJ*, Vol. LXVIII (January 7, 1899), 20–21.
71. "The Joplin Zinc Ore Market," *EMJ*, Vol. LXVII (February 4, 1899), 138.
72. "Mining News," *EMJ*, Vol. LII (August 22, 1891), 224, and "Mining News," *EMJ*, Vol. LII (December 5, 1891), 468.
73. Nicholson, "The Missouri-Kansas Zinc District," *EMJ*, Vol. LXVII (June 10, 1899), 680.
74. Zook, "Conditions in the Joplin District," *EMJ*, Vol. LXXIX (January 12, 1905), 86.
75. "Editorial Correspondence," *EMJ*, Vol. CII (June 16, 1917), 1096.
76. Tri-State Survey Committee, *A Preliminary Report on Conditions in the Tri-State Area*, 12.
77. "Mining News," *EMJ*, Vol. CXX (August 22, 1925), 304.
78. "Cooperative Work," *American Mining Congress and American Institute of Mining and Metallurgical Engineers Convention Booklet* (September 28, 1931), 26.
79. "Washington Letter," *EMJ*, Vol. CXXIII (May 14, 1927), 816.
80. "Cooperative Work," *American Mining Congress and American Institute of Mining and Metallurgical Engineers Convention Booklet* (September 28, 1931), 26.
81. "Mining News," *EMJ*, Vol. XLIX (April 12, 1890), 429.
82. "Mining News," *EMJ*, Vol. LXXIV (October 19, 1907), 756.
83. Crane and Adams, "Lead and Zinc Mining In the Quapaw District," *Mines and Minerals* (May, 1907), 445–46.
84. *Daily Oklahoman*, May 12, 1953.
85. *Problems in the Metal Mining Industry*, 515.
86. Ernest Gent, *The Zinc Industry, A Mine to Market Outline*, 1949.

XII. TRI-STATE DISTRICT SOCIAL CONDITIONS

1. Martin, *Statistics of Production of Lead and Zinc*, 16.
2. O. N. Wampler, "Safety and Industrial Relations at Eagle Picher," *Mining Congress Journal*, Vol. XV (November, 1929), 876.
3. Missouri Bureau of Mines, *Fifteenth Annual Report*; Missouri Bureau of Mines, *Seventeenth Annual Report*; Haworth, Crane, and Rogers, *Special Report*; and Oklahoma Bureau of Mines, *First Annual Report*.
4. Keener, *Method and Cost at Hartley-Grantham Mine*, 6.
5. Wright, *Mining and Milling*, 121.
6. Oklahoma Bureau of Mines, *First Annual Report*, 73.
7. Missouri Bureau of Mines, *Seventeenth Annual Report*, 21.
8. "Imprisoned in a Joplin Mine," *EMJ*, Vol. XCIII (August 5, 1911), 252.

9. Snider, *Preliminary Report*, 73.
10. Wright, *Mining and Treatment*, 39.
11. Wright, *Mining and Milling*, 121.
12. S. P. Howell, "Prevention of Accidents with Explosives," *American Zinc, Lead, and Copper Journal*, Vol. XX (June, 1928), 4.
13. Missouri Bureau of Mines, *Seventeenth Annual Report*, 21–22.
14. Wright, *Mining and Milling*, 15.
15. "Cooperative Work: Tri-State Ore Producers' Association," *American Mining Congress and American Institute of Mining and Metallurgical Engineers Convention Booklet* (September 28, 1931), 26.
16. Wampler, "Safety and Industrial Relations at Eagle Picher," *Mining Congress Journal*, Vol. XV (November, 1929), 878.
17. Giessing, "Accident Prevention," *Mining Congress Journal*, Vol. XV (November, 1929), 853.
18. See Herbert Hoover and L. H. Hoover, trans., *Georgius Agricola, de re Metallica.*
19. Wright, *Mining and Treatment*, 40.
20. "Editorial Correspondence," *EMJ*, Vol. XCVIII (December 12, 1914), 1062.
21. "Editorial Correspondence," *EMJ*, Vol. XCVII (March 12, 1914), 536.
22. "Editorial Correspondence," *EMJ*, Vol. XCVII (May 2, 1914), 924.
23. "Editorial Correspondence," *EMJ*, Vol. XCVIII (November 14, 1914), 983.
24. "Tuberculosis in the Joplin District," *EMJ*, Vol. XCIX (February 13, 1915), 331.
25. Higgins et al., *Siliceous Dust*, 67.
26. Anthony J. Lanza, *A Preliminary Report on Sanitary Conditions in the Zinc-Lead Mines of the Joplin, Missouri District and Their Relation to Pulmonary Diseases Among the Miners*, 5–7.
27. "Editorial Correspondence," *EMJ*, Vol. CII (December 30, 1916), 1153.
28. Higgins et al., *Siliceous Dust*, 20–21.
29. Lallah S. Davidson, *South of Joplin: Story of a Tri-State Diggins'*, 10; Tri-State Survey Committee, *A Preliminary Report on Living, Working, and Health Conditions in the Tri-State Mining Area*, 82; and Frederick L. Ryan, *Problems of the Oklahoma Labor Market*, 30.
30. Higgins et al., *Siliceous Dust*, 65.
31. *Ibid.*, 70–73.
32. *Revised Statutes of the State of Missouri, 1939*, II, 3792.
33. *Kansas Mining Law, General Statutes of Kansas, 1949*, I, 252. Also see *Mining Code, Oklahoma Statutes Annotated*, 622–36.
34. Wright, *Mining and Milling*, 27.
35. R. R. Sayers et al., *Silicosis and Tuberculosis Among Miners of*

the Tri-State District of Oklahoma, Kansas, and Missouri, U.S. Bureau of Mines *Technical Paper* No. 645, 1–3.

36. *Ibid.*, 6–7.

37. *Ibid.*, 5.

38. *Ibid.*, 6–7.

39. F. V. Meriwether, R. R. Sayers, and A. J. Lanza, *Silicosis and Tuberculosis Among Miners of the Tri-State District of Oklahoma-Kansas-Missouri for the Year Ended June 30, 1929*, U.S. Bureau of Mines *Technical Paper* No. 552, 5.

40. *Ibid.*

41. *Ibid.*, 6.

42. *Ibid.*, 27.

43. H. C. Chellson, "Dust Count Technique in Tri-State Mines," *EMJ*, Vol. CXL (December, 1939), 30.

44. Frances Perkins, *Conference on Health and Working Conditions in the Tri-State District, Joplin, Missouri, 1940*, 35.

45. Meriwether, Sayers, and Lanza, *Silicosis and Tuberculosis*, 27.

46. "Cooperative Work," *American Mining Congress and American Institute of Mining and Metallurgical Engineers Convention Booklet* (September 28, 1931), 26.

47. Alice Hamilton, "A Mid-American Tragedy," *Survey Graphic*, Vol. XXIX (August, 1940), 435.

48. Meriwether, Sayers, and Lanza, *Silicosis and Tuberculosis*, 27.

49. Frederick L. Ryan, *A History of Labor Legislation in Oklahoma*, 76.

50. "Lead Poisoning in the Mining of Lead," *Mining Congress Journal*, Vol. XIII (March, 1927), 245.

51. Higgins et al., *Siliceous Dust*, 73.

52. "Mining News," *EMJ*, Vol. XXIV (July 21, 1877), 49.

53. "Mining News," *EMJ*, Vol. XXIV (July 28, 1877), 71.

54. "Prevention of Lead Poisoning," *EMJ*, Vol. XXVI (September 14, 1878), 184.

55. Alice Hamilton, *Lead Poisoning in the Smelting and Refining of Lead*, 141.

56. *Norman Transcript*, October 22, 1948.

57. Jasper County Health Department, "Annual Report of the Jasper County Health Department" (Webb City, 1940). Mimeographed. Oklahoma Health Department, "Tuberculosis in Oklahoma" (Oklahoma City, 1945). Mimeographed. Marian Friedman, "Annual Report of the Cherokee County Health Department," prepared for the Cherokee County Health Department (Columbus, 1944). Mimeographed.

58. Perkins, *Conference on Health and Working Conditions*, 34.

59. "Cooperative Work," *American Mining Congress and American Institute of Mining and Metallurgical Engineers Convention Booklet* (September 28, 1931), 26.

60. Perkins, *Conference on Health and Working Conditions*, 14.

61. *Daily Oklahoman*, November 21, 1948.

62. "Health Status of the People of One Kansas County," speech delivered by Joseph W. Spearing before the Kansas Nutrition Conference (Topeka, October 17, 1941). Mimeographed.

63. Cleo Stricklin, "Tuberculosis in Cherokee County, Kansas, Special Report for the Kansas Tuberculosis and Health Association and the Public Health Department, Cherokee County, Kansas," prepared for the Kansas State Department of Health (Topeka, 1953), 6. Mimeographed.

64. "Tuberculosis in Oklahoma," *Oklahoma Health Bulletin*, Vol. X (November, 1952), 44.

65. See Chapter XIV.

66. Martin, *Statistics of Production of Lead and Zinc*, 22.

67. *Eagle Picher and International Union*, 760, 767.

68. Clifton Hall, "A Report of Tuberculin Testing in Cherokee County," prepared for the Division of Tuberculosis Control, Kansas State Board of Health (Topeka, 1936). Mimeographed.

69. *Daily Oklahoman*, November 21, 1948.

XIII. TRI-STATE LABOR CONDITIONS: THE EARLY PERIOD

1. Data for these mining employee designations have been extracted from U.S. Bureau of Labor Statistics, *Wages and Hours of Labor in Metalliferous Mines, 1924 and 1931, Bulletin* No. 573, 5; U.S. Bureau of Labor Statistics, *Wage Structure of the Nonferrous Metals Industry, 1941–1942, Bulletin* No. 729, 14–15; and U.S. Bureau of Labor Statistics, *Wages in Lead and Zinc Mines and Mills in the Tri-State Area, June, 1943, Bulletin* No. 1578, 5.

2. Broadhead, *Report, 1873–1874*, 491–92.

3. Shaner, *The Story of Joplin*, 10.

4. Missouri Bureau of Labor Statistics, *Eleventh Annual Report*, 336.

5. James L. Bruce, "Ore Dressing in the Joplin District," *EMJ*, Vol. XCIII (February 24, 1912), 405.

6. Data for the employment census have been extracted from Ryan, *Problems of the Oklahoma Labor Market*, 98; U.S. Bureau of Labor Statistics publications: *Wages and Hours of Labor, 1924 and 1931*, 3; *Wage Structure of the Nonferrous Metals Industry, 1941–1942*, 4; *Wages in the Nonferrous Metals Industry, June, 1943*, 6; *Wages in Lead and Zinc Mines and Mills, 1943*, 103; and U.S. Congress, House, Hearings Before the Select Committee on Small Business, *Problems in the Metal-Mining Industry*, 83 Cong., 1 sess., 1953, 443–45.

7. Missouri Bureau of Labor Statistics, *Ninth Annual Report*, 255.

8. Charles P. Williams, *Industrial Report on Lead, Zinc, and Iron*, 44.

9. *Mining Code, Revised Statutes of Missouri, 1899*, II, 2042.

10. "The Eight-Hour Law in Missouri," *EMJ*, Vol. LXXX (December 2, 1905), 1026.

11. *Oklahoma Constitution*, Article XXIII, Sections 1 and 4.

12. *Kansas Mining Law, General Statutes of Kansas Annotated*, 282.

13. *United States Statutes at Large*, XLVIII, 199.

14. *Ibid.*, LII, 1063. The power of the states to regulate hours in mines was upheld by the United States Supreme Court in 1898. See *Holden* v. *Hardy*, *United States Reports*, CLXIX, 366–98. This involved the validity of a Utah eight-hour law.

15. *Daily Oklahoman*, December 17, 1922.

16. Data for wages were extracted from the following sources: Missouri Bureau of Labor Statistics, *Ninth Annual Report*, 255; Missouri Bureau of Labor Statistics, *Eleventh Annual Report*, 336; Walter R. Ingalls, "Missouri and Kansas Zinc Production," *EMJ*, Vol. LXXXIII (May 18, 1907), 939; "Editorial Correspondence," *EMJ*, Vol. CII (September 2, 1916), 442; Kitson, "The Mining Districts of Joplin and Southeast Missouri," *EMJ*, Vol. CV (March 2, 1918), 415; "Editorial Correspondence," *EMJ*, Vol. CVIII (August 2, 1919), 199; Wright, *Mining and Treatment*, 36–37; Wright, *Mining and Milling*, 117; U.S. Bureau of Labor Statistics, *Wages and Hours of Labor in Metalliferous Mines, 1924 and 1931*, 8–9; Keener, *Method and Cost at Barr Mine*, 5; Keener, *Method and Cost at Hartley-Grantham Mine*, 4; *Wages in Lead and Zinc Mines and Mills*, 4–5; and U.S. Bureau of Labor Statistics, *Hours and Earnings in Lead and Zinc Mining*, 1.

17. "Editorial Correspondence," *EMJ*, Vol. CI (April 22, 1916), 751.

18. Ruhl, "Keeping Mining Costs at Joplin, *EMJ*, Vol. XCII (December 9, 1911), 1134.

19. See Chapter VII for an account of modern mining methods applied in the Tri-State District since 1930.

20. Ross, *Death of a Yale Man*, 185.

21. Britton, *Pioneer Life*, 20.

22. Burgess, "Mining Costs in the Missouri-Kansas Districts," *Mining and Engineering World*, Vol. XXXVIII (April, 1913), 804.

23. *Ibid.*

24. Williams, *The State of Missouri*, 294.

25. Carter, "Economic Conditions in the Joplin District," *EMJ*, Vol. XC (October 15, 1910), 760.

26. A. Heckscher, "Zinc and Zinc Oxide in 1898," *EMJ*, Vol. LXVIII (January 7, 1899), 20.

27. Wright, *Mining and Treatment*, 36.

28. *Daily Oklahoman*, December 17, 1922.

29. Zook, "Zinc and Lead in the Joplin District in 1908," *EMJ*, Vol. LXXXVII (January 9, 1909), 71.

30. "Mining News," *EMJ*, Vol. XC (July 6, 1910), 234.

31. Ruhl, "Keeping Mining Costs at Joplin," *EMJ*, Vol. XCII (December 9, 1911), 1134.

32. Otto Ruhl, "Zinc in Missouri, Kansas, and Oklahoma," *EMJ*, Vol. CLXVI (August 17, 1918), 302.

33. J. B. Rhyne, *Social and Community Problems of Oklahoma*, 111.

34. U.S. Bureau of the Census, *Thirteenth Census of the United States: 1910, Population*, III, 461, and *Seventeenth Census of the United States: 1950, Population*, I, 69, 89, 130.

35. *Thirteenth Census of the United States*, II, 676, and *Seventeenth Census of the United States*, I, 91.

36. Henry J. Allen, *The Party of the Third Part: The Story of the Kansas Industrial Relations Court*, 49.

37. "Mining News," *EMJ*, Vol. CXIII (April 22, 1922), 690.

38. Perkins, *Conference on Health and Working Conditions*, 9.

39. "Editorial Correspondence," *EMJ*, Vol. C (August 14, 1915), 288.

40. Missouri Bureau of Labor Statistics, *Ninth Annual Report*, 255.

41. Carter, "Economic Conditions in the Joplin District," *EMJ*, Vol. XC (October 15, 1910), 760.

42. Stone, "The Lead and Zinc Field of Kansas," *Kansas Historical Collections*, Vol. VII (1902), 255–56.

43. Carthage *Weekly Banner*, March 7, 1872.

44. Livingston, *Jasper County*, I, 57.

45. Carthage *Weekly Banner*, March 7, 1872.

46. Granby *News Herald*, Centennial Edition, June 1, 1950.

47. North, *Jasper County*, 401.

48. *Ibid.*, 244.

49. Livingston, *Jasper County*, I, 178.

50. Broadhead, *Report, 1873–1874*, 490.

51. Missouri Bureau of Mines, *Fifteenth Annual Report*, 18–19.

52. Granby *News Herald*, Centennial Edition, June 1, 1950.

53. Draper, *Old Grubstake Days in Joplin*, 18–19.

54. Carthage *Weekly Banner*, August 10, 1871.

55. Missouri Bureau of Mines, *Eleventh Annual Report*, 336–37.

XIV. TRI-STATE LABOR CONDITIONS: THE RECENT PERIOD

1. Missouri Bureau of Mines, *Seventeenth Annual Report*, 31.

2. Bruce, "Ore Dressing in the Joplin District," *EMJ*, Vol. XCIII (February 24, 1912), 407.

3. "Editorial Correspondence," *EMJ*, Vol. XCVIII (February 21, 1914), 446.

4. "Labor Troubles in Mining," *EMJ*, Vol. C (July 10, 1915), 78.

5. *Mining Code, Revised Statutes of Missouri, 1899*, II, 2042.

6. "Labor Troubles at Joplin," *EMJ*, Vol. LXXX (December 16, 1905), 1121.

7. "The Joplin District," *EMJ*, Vol. LXXXI (February 3, 1906), 234–35.

8. Carl Junction *Socialist News*, October 2, 1906.

9. Missouri Bureau of Labor Statistics, *Eleventh Annual Report*, 340.

10. "Mining News," *EMJ*, Vol. LVI (August 12, 1893), 172.
11. Ryan, *A History of Labor Legislation In Oklahoma*, 72, 103.
12. *Missouri Trades Unionist*, May 31, 1911.
13. "Mining Conditions in the Joplin District," *EMJ*, Vol. LXXXIX (June 4, 1910), 1144.
14. Ruhl, "Keeping Mining Costs at Joplin," *EMJ*, Vol. XCII (December 9, 1911), 1134.
15. "Editorial Correspondence," *EMJ*, Vol. XCVII (February 21, 1914), 446.
16. "Editorial Correspondence," *EMJ*, Vol. C (July 10, 1915), 82.
17. "Editorial Correspondence," *EMJ*, Vol. C (July 17, 1915), 122.
18. *Missouri Trades Unionist*, February 28, 1916.
19. Ryan, *A History of Labor Legislation in Oklahoma*, 72, 103.
20. *Missouri Trades Unionist*, May 31, 1911.
21. Carl Junction *Socialist News*, October 2, 1906.
22. *Ibid.*, January 4, 1906.
23. *Ibid.*, October 2, 1906.
24. *Ibid.*, December 18, 1906.
25. *Ibid.*, October 21, 1907.
26. *Ibid.*, January 22, 1907.
27. Ruhl, "Keeping Mining Costs at Joplin," *EMJ*, Vol. XCII (December 9, 1911), 1134.
28. "Labor Troubles at Joplin," *EMJ*, Vol. LXXX (December 16, 1905), 77–78.
29. Joplin *Globe*, July 6, 1939, and Joplin *News Herald*, September 15, 1939.
30. *United States Statutes at Large*, XLVIII, 198.
31. "Mining News," *EMJ*, Vol. CXXXVI (July, 1935), 347.
32. Extracted from United States Bureau of the Census, *Fifteenth Census of the United States: 1930, Unemployment*, I, 384, 576, 819.
33. Ross, *Death of a Yale Man*, 185.
34. Ryan, *Problems of the Oklahoma Labor Market*, 98.
35. Martin, *Statistics of Production of Lead and Zinc*, 21.
36. Joplin *Globe*, May 9, 1935.
37. Ryan, *Problems of the Oklahoma Labor Market*, 31.
38. Miami *Daily Record*, May 21, 1935.
39. *Eagle Picher Mining and Smelting Company and International Union of Mine, Mill, and Smelter Workers, Locals Nos. 15, 17, 107, 108, and 111, Decisions and Orders of the National Labor Relations Board, October 16–31, 1939*, XVI, 741. Hereafter cited as *Eagle Picher and International Union.*
40. Miami *Daily Record*, May 21, 1935.
41. Albert G. Pickerell, "Lead Miners in Missouri," *New Republic*, Vol. XCII (August 11, 1937), 12–13.
42. *Eagle Picher and International Union*, 750.
43. Ross, *Death of a Yale Man*, 193.

44. *Daily Oklahoman*, October 13, 1950.
45. *Eagle Picher and International Union*, 743.
46. Miami *Daily Record*, May 26, 1935.
47. "Mining News," *EMJ*, Vol. CXXXVI (July, 1935), 317.
48. *Eagle Picher and International Union*, 745.
49. Miami *Daily Record*, May 30, 1935.
50. *Ibid.*, June 2, 1935.
51. *Ibid.*, June 9, 1953.
52. *Eagle Picher and International Union*, 749.
53. *Ibid.*, 746–47.
54. *Ibid.*, 762.
55. *Ibid.*, 768.
56. Miami *Daily Record*, May 30, 1935.
57. *Eagle Picher and International Union*, 754.
58. *Ibid.*
59. Harry A. Millis and Emily Clark Brown, *From the Wagner Act to Taft-Hartley*, 100.
60. The rustling card was a personnel device used in the district to indicate that an applicant was satisfactory for employment. This was generally obtained from the ground boss and consisted of an informal note stating that he was willing to have the applicant assigned to his crew.
61. Unions customarily issued membership credentials to members in good standing. In most cases this was a small booklet containing the member's identification, credit for union dues paid, and several blank pages. Whenever the union went out on strike, members were required to take their turn at picketing. Credit for picketing service was indicated by union officers pasting small stamps on the blank pages of the membership booklets. While in respectable union circles picket stamps served as a mark of prestige and esteem, to the Blue Card Union, they were a badge of dishonor and furnished a reason for refusing employment.
62. *Eagle Picher and International Union*, 759.
63. *Ibid.*, 746.
64. Davidson, *South of Joplin*, 270.
65. *Eagle Picher and International Union*, 757–58.
66. *United States Statutes at Large*, XLIX, 449.
67. *Eagle Picher and International Union*, 760, 767.
68. *National Labor Relations Board* v. *Jones and Laughlin Steel Corporation*, *United States Reports*, CIII, 1–49.
69. *Eagle Picher and International Union*, 758.
70. *Ibid.*, 771.
71. *United States Statutes at Large*, XLIX, 453.
72. *Daniel M. Lyons et al.* v. *the Eagle Picher Company*, *Federal Reporter*, *Second Series*, XC, 321–23.

73. *Eagle Picher and International Union*, 728.

74. *Ibid.*, 766.

75. *Ibid.*, 874–82.

76. *United States Statutes at Large*, XLIX, 449.

77. *Eagle Picher Mining and Smelting Company* v. *National Labor Relations Board, Federal Reporter, Second Series*, CXIX, 903.

78. *Ibid.*, 907.

79. *In the Matter of the Tri-State Zinc Company and International Union of Mine, Mill, and Smelter Workers, Affiliated with the C.I.O., Decisions and Orders of the National Labor Relations Board*, XXXIX, 1095.

80. Perkins, *Conference on Health and Working Conditions*, 35.

81. *Ibid.*, 36.

82. U.S. Bureau of Labor Statistics, *Wage Structure of the Nonferrous Metals Industry, 1941–1942*, 5–6.

83. *Ibid.*, 6.

84. U.S. Bureau of Labor Statistics, *Wages in the Nonferrous Metals Industry, June, 1943*, 7.

85. U.S. Bureau of Labor Statistics, *Wages in Lead and Zinc Mines and Mills in the Tri-State Area, June, 1943*, 1.

86. Joplin *Globe*, March 26, 1949.

87. *Ibid.*, August 1, 1948.

88. *Ibid.*

89. *Daily Oklahoman*, March 31, 1953.

90. Joplin *Globe*, April 2, 1953.

91. *Ibid.*, June 20, 1953.

92. Miami *Daily Record*, June 21, 1953.

93. *Daily Oklahoman*, October 29, 1953.

94. *Ibid.*, December 22, 1953.

95. Miami *Daily Record*, July 22, 1953.

XV. TRI-STATE SOCIETY

1. Anthony J. Lanza, *Miners Consumption: A Study of 433 Cases of the Disease Among Zinc Miners in Southwest Missouri*, U.S. Public Health Service *Bulletin* No. 85, and Higgins et al, *Siliceous Dust.*

2. Sayers et al., *Silicosis and Tuberculosis Among Miners of the Tri-State District*, and Meriwether, Sayers, and Lanza, *Silicosis and Tuberculosis.*

3. Livingston, *Jasper County*, I, 150.

4. Granby *Miner*, October 4, 1873.

5. Missouri Bureau of Labor Statistics, *Ninth Annual Report*, 255.

6. Granby *Miner*, October 4, 1873.

7. Burgess, "Mining Costs in the Missouri-Kansas District," *Mining and Engineering World*, Vol. XXXVIII (April, 1913), 804.

8. Draper, *Though Long the Trail*, 171.

9. Walter Williams, *The State of Missouri*, 294.

10. Carter, "Economic Conditions in the Joplin District," *EMJ*, Vol. XC (October 15, 1910), 759.
11. North, *Jasper County*, 396.
12. *Daily Oklahoman*, August 12, 1917.
13. Broadhead, "Southwest Missouri Lead Interests," *EMJ*, Vol. XXV (February, 1883), 73.
14. Missouri Bureau of Labor Statistics, *Eleventh Annual Report*, 378–79.
15. Holibaugh, *The Lead and Zinc Mining Industry*, 13.
16. Rhyne, *Social and Community Problems of Oklahoma*, 111.
17. "The Missouri Lead and Zinc Company's Plant," *EMJ*, Vol. LXIX (June 2, 1900), 648.
18. Picher *Tri-State Tribune*, December 21, 1951.
19. Missouri Bureau of Labor Statistics, *Eleventh Annual Report*, 337.
20. Lanza, *Miners Consumption*, 337.
21. *Ibid.*, 76.
22. *Ibid.*, 182.
23. Perkins, *Conference on Health and Working Conditions*, 35–36.
24. Extracts from letter of Kansas Secretary of State, Topeka, December 23, 1953, letter of Missouri Secretary of State, Jefferson City, January 4, 1954, and *Directory of the State of Oklahoma*, 1951.
25. U.S. Bureau of the Census, *Special Reports, Religious Bodies, 1906*, I; *Religious Bodies, 1916*, I; *Religious Bodies, 1926*, I; and *Religious Bodies of the United States, 1936*, I. These have served as sources for the foregoing data on Tri-State religious life.
26. Draper, *Though Long the Trail*, 172.
27. *Illustrated Historical Atlas*, 19.
28. Per capita statistics extracted from letter of Department of Education, State of Missouri, Jefferson City, October 8, 1953; letter of State Board of Education, State of Oklahoma, Oklahoma City, October 15, 1953; and letter of State Department of Education, State of Kansas, Topeka, October 25, 1953.
29. Pupil loss statistics derived from U.S. Bureau of the Census, *Thirteenth Census of the United States: 1910, Population*, II; *Fourteenth Census of the United States: 1920, Population*, II; and *Fifteenth Census of the United States: 1930, Population*, II.
30. Bruce, "Ore Dressing in the Joplin District," *EMJ*, Vol. XCIII (February 24, 1912), 405.
31. Missouri Bureau of Labor Statistics, *Eleventh Annual Report*, 336–37.
32. Britton, *Pioneer Life*, 19.
33. Draper, *Old Grubstake Days in Joplin*, 18–20.
34. "Mining News," *EMJ*, Vol. XLIX (April 19, 1890), 454.
35. Missouri Bureau of Labor Statistics, *Eleventh Annual Report*, 338–46.

36. U.S. Bureau of the Census, *Eighth Census of the United States: 1860, Population*, I; *Ninth Census of the United States: 1870, Population*, I; *Tenth Census of the United States: 1880, Population*, I; *Eleventh Census of the United States: 1890, Population*, II; *Twelfth Census of the United States: 1900, Population*, I; *Thirteenth Census of the United States: 1910, Population*, II; *Fourteenth Census of the United States: 1920, Population*, I; *Fifteenth Census of the United States: 1930, Population*, II; *Sixteenth Census of the United States: 1940, Population*, II; and *Seventeenth Census of the United States: 1950, Population*, I.

37. Lanza, *Miners Consumption*, 15.

38. Granby *Miner*, October 11, 1873.

39. Missouri Bureau of Labor Statistics, *Eleventh Annual Report*, 378–89.

40. Oklahoma Bureau of Mines, *First Annual Report*, 37.

41. "Why Mining at Joplin Does Not Languish," *EMJ*, Vol. XCVII (January 24, 1914), 232–33.

42. "Mining News," *EMJ*, Vol. CXXII (November 12, 1921), 793.

43. "Mining News," *EMJ*, Vol. CXXXI (May 31, 1931), 434.

44. Ross, *Death of a Yale Man*, 185.

XVI. THE TRI-STATE DISTRICT AT MID-CENTURY

1. Ruhl, Allen, and Holt, *Zinc-Lead Ore Reserves of the Tri-State District*, 16.

2. Tri-State Ore Producers' Association, "Why Conservation of Marginal Zinc and Lead Ores Means So Much to Our National Economy and Security" (Picher, 1946), 6. Mimeographed.

3. Ed Edmondson, "Preliminary Report on Zinc in Oklahoma," prepared by Representative Edmondson, Second District, Oklahoma, for the Congress of the United States, House of Representatives (Washington, 1953), 1–5. Mimeographed.

4. Tri-State Ore Producers' Association, "Tri-State District Lead and Zinc Mines Report" (Picher, 1954), 1. Mimeographed.

5. *Daily Oklahoman*, April 5, 1953.

6. Joplin *Globe*, April 3, 1953.

7. Edmondson, "Preliminary Report on Zinc in Oklahoma" (Washington, 1953), 3. Mimeographed.

8. U.S. Congress, House, Hearings Before the Select Committee on Small Business, *Problems in the Metal Mining Industry (Lead, Zinc, and Other Metals)*, 83 Cong., 1 sess., 1953.

9. U.S. Congress, House, *House Bill No. 4294*, 83 Cong., 1 sess., 1953, and Senate, *Senate Bill No. 869*, 83 Cong., 1 sess., 1953.

10. U.S. Congress, *Congressional Record*, 83 Cong., 2 sess., 1954, C, 726.

11. *Ibid.*

12. Tri-State Ore Producers' Association, "Conservation of Marginal Zinc" (1946), 6. Mimeographed.

Selected Bibliography

I. *Bibliographies*

Annual American Catalog, 1886–1900. New York, Publishers Weekly Co., 1887–1901.

Annual American Catalog, 1900–1909. 7 vols. New York, Publishers Weekly Co., 1900–1909.

Bain, Foster H. *Index to Bureau of Mines Publications.* Washington, Government Printing Office, 1922.

"Bibliography of Kansas Geology," *Transactions of the Kansas Academy of Science,* Vol. XIV (1896), 261–78.

Breger, C. L. and G. E. Sisley. *The Mining World Index of Current Literature, 1912–1916.* Chicago, Mining World Company, 1916.

Butcher, George M., et al. *A Guide to Historical Literature,* New York, MacMillan Co., 1931.

Carroll, H. C. *Index to Bureau of Mines Papers Published in the Technical Press, 1910 to 1930.* Washington, Government Printing Office, 1931.

Cushing, H. G., and A. V. Morris. *Nineteenth Century Reader's Guide to Periodical Literature.* 2 vols. New York, H. W. Wilson Co., 1944.

Faxon, Frederick W. *Magazine Subject Index.* London, Library Association Publication, 1919.

Fletcher, William F., and Mary Poole. *Poole's Index to Periodical Literature, 1815–1899.* Boston, Houghton Mifflin Co., 1907.

Gregor, Darling R. *Bibliography of the Geology of Missouri.* Jefferson City, Mid-State Printing Co., n.d.

Gregory, Winifred. *Union List of Serials in Libraries of the*

United States and Canada. New York, H. W. Wilson Co., 1927.

Griffin, Appleton. *Bibliography of American Historical Societies.* Washington, Government Printing Office, 1907.

Hargrett, Lester. *A Bibliography of the Constitution and Laws of the American Indian.* Cambridge, Harvard University, 1947.

———. *Oklahoma Imprints, 1835–1890.* New York, R. R. Bowker Co., 1951.

Larned, Joseph N. *The Literature of American History: A Bibliography.* Boston, American Library Association Publication, 1902.

Logasa, Hannah. *Regional United States.* Boston, F. W. Faxon Co., 1942.

Nickles, John M. *Bibliography and Index of Geology Exclusively of North America.* New York, Geological Society of America, 1933.

———. *Geological Literature of North America, 1785–1928.* 3 vols. Washington, U.S. Geological Survey, 1923–31.

Poore, Ben P. *Descriptive Catalog of the Government Publications of the United States, 1774–1881.* Washington, Government Printing Office, 1901.

Potter, Manion. *The United States Catalog.* Minneapolis, H. W. Wilson Co., 1903.

———. *The United States Catalog.* Minneapolis, H. W. Wilson Co., 1912.

Rader, Jesse L. *South of Forty: From the Mississippi to the Río Grande, A Bibliography.* Norman, University of Oklahoma Press, 1947.

Sabine, Joseph. *Dictionary of Books Relating to America.* New York, Bibliographical Society of America, 1892.

Superintendent of Documents. *Checklist of United States Publications, 1789–1909.* Washington, Government Printing Office, 1911.

Superintendent of Documents. *Catalog of Public Documents of the Congress and All Departments of the Government of the United States.* Washington, Government Printing Office, 1893.

Thom, Emma. *Bibliography of North American Geology, 1929–1939.* U.S. Geological Survey *Bulletin* No. 397. Washington, Government Printing Office, 1944.

———. *Bibliography of North American Geology, 1933–1934.* U.S. Geological Survey *Bulletin* No. 869. Washington, Government Printing Office, 1935.

———. *Bibliography of North American Geology, 1942.* U.S. Geological Survey *Bulletin* No. 949. Washington, Government Printing Office, 1945.

———. *Bibliography of North American Geology, 1944–1945.* U.S. Geological Survey *Bulletin* No. 952. Washington, Government Printing Office, 1947.

———. *Bibliography of North American Geology, 1946–1947.* U.S. Geological Survey *Bulletin* No. 958. Washington, Government Printing Office, 1949.

———. *Bibliography of North American Geology, 1948.* U.S. Geological Survey *Bulletin* No. 968. Washington, Government Printing Office, 1949.

Thom, Emma, et al. *Bibliography of North American Geology, 1949.* U.S. Geological Survey *Bulletin* No. 977. Washington, Government Printing Office, 1951.

U.S. Bureau of the Census. *Circular of Information Covering Census Publications, 1790–1916.* Washington, Government Printing Office, 1917.

Winslow, Mabel. *Bureau of Mines Papers Published in the Technical Press, January 1, 1938 to January 1, 1946.* Washington, U.S. Bureau of Mines, 1946.

Works Progress Administration. *Historical Records Survey: American Imprints Inventory.* Washington, Government Printing Office, 1937.

II. *Federal Documents*

Annual Report of the Commissioner of Indian Affairs to the Secretary of the Interior for the Year 1891. Washington, Government Printing Office, 1891.

Bain, H. F., C. R. Van Hise, and G. I. Adams. *Preliminary Report on the Lead and Zinc Deposits of the Ozark Region.*

U.S. Geological Survey, Twenty-second *Annual Report.* Washington, Government Printing Office, 1901.

Banks, L. M. *Mining Methods and Costs in the Waco District.* U.S. Bureau of Mines *Information Circular* No. 6150. Washington, Government Printing Office, 1929.

Brichta, Louis C. *Report of Exploration for Zinc and Lead Ore, Phelps Lease, Jasper County, Missouri.* U.S. Bureau of Mines *Report of Investigations* No. 3941. Washington, Government Printing Office, 1946.

———. *Report of Exploration of O'Jack Mining Company Zinc and Lead Deposits, Jasper County, Missouri.* U.S. Bureau of Mines *Report of Investigations* No. 3970. Washington, Government Printing Office, 1946.

Coghill, W. H., and Carl O. Anderson. *Milling Methods in the Tri-State District.* U.S. Bureau of Mines *Report of Investigations* No. 2314. Washington, Government Printing Office, 1922.

Crabtree, Edwin H. *Milling Practice at the White Bird Concentrator.* U.S. Bureau of Mines *Report of Investigations* No. 6355. Washington, Government Printing Office, 1930.

Daniel M. Lyons et al. v. *Eagle Picher Co. Federal Reporter,* Second Series. XC. St. Paul, West Publishing Co., 1937.

Eagle-Picher Mining and Smelting Company and International Union of Mine, Mill, and Smelter Workers, Locals Nos. 15, 17, 107, 108, and 111. Decisions and Orders of the National Labor Relations Board, October 16–31, 1939, XVI. Washington, Government Printing Office, 1940.

Eagle-Picher Mining and Smelting Company et al. v. *National Labor Relations Board (International Union of Mine, Mill, and Smelter Workers, Locals Nos. 15, 17, 107, 108, and 111.) Federal Reporter,* Second Series. CXIX. St. Paul, West Publishing Co., 1941.

Hamilton, Alice. *Lead Poisoning in the Smelting and Refining of Lead.* Washington, U.S. Department of Labor, 1914.

Higgins, Edwin et al. *Siliceous Dust in Relation to Pulmonary Disease Among Miners in the Joplin District, Missouri.* U.S. Bureau of Mines *Bulletin* No. 132. Washington, Government Printing Office, 1917.

Holden v. *Hardy*. *United States Reports*. New York, n.p., 1898.

Hornor, R. R., and H. E. Tufft. *Mine Timber, Its Selection, Storage, Treatment, and Use*. Washington, U.S. Bureau of Mines, 1925.

———. *Pillar and Room Methods of the Joplin District*. U.S. Bureau of Mines *Bulletin* No. 235. Washington, Government Printing Office, 1925.

Jackson, Charles F., et al., *Lead and Zinc Mining and Milling in the United States: Current Practices and Costs*. U.S. Bureau of Mines *Bulletin* No. 381. Washington, Government Printing Office, 1935.

Jenny, W. F. *A Preliminary Report of the Work on the Lead and Zinc Fields of Missouri*. U.S. Geological Survey *Eleventh Annual Report*. Washington, Government Printing Office, 1891.

Keener, Oliver W. *Method and Cost at Barr Mine, Tri-State District*. U.S. Bureau of Mines *Information Circular* No. 6159. Washington, Government Printing Office, 1929.

———. *Method and Cost at Hartley-Granthem Mine, Tri-State Zinc and Lead District*. U.S. Bureau of Mines *Information Circular* No. 6286. Washington, Government Printing Office, 1930.

Lanza, Anthony J. *A Preliminary Report on Sanitary Conditions in the Zinc-Lead Mines of the Joplin, Missouri District and Their Relation to Pulmonary Diseases Among the Miners*. Washington, U.S. Bureau of Mines, 1915.

———. *Miners Consumption: A Study of 433 Cases of the Disease Among Zinc Miners in Southwest Missouri*. U.S. Public Health Service *Bulletin* No. 85. Washington, Government Printing Office, 1917.

Lowrie, A. L., et al., eds. *The American State Papers: Public Lands*, I, II, VIII. Washington, Green and Co. and Gales and Seaton, 1832–61.

Martin, A. J. *Summarized Statistics of Production of Lead and Zinc in the Tri-State Mining District*. U.S. Bureau of Mines *Information Circular* No. 7383. Washington, Government Printing Office, 1946.

Meriwether, F. V., R. R. Sayers, and A. J. Lanza. *Silicosis and Tuberculosis Among Miners of the Tri-State District of Oklahoma-Kansas for the Year Ended June 30, 1929.* U.S. Bureau of Mines *Technical Paper* No. 552. Washington, Government Printing Office, 1929.

Moon, Lowell B. *Bureau of Mines Strategic Minerals Development Program.* U.S. Bureau of Mines *Report of Investigations* No. 4647. Washington, Government Printing Office, 1950.

National Labor Relations Board v. *Jones and Laughlin Steel Corporation. United States Reports.* CIII. Washington, Government Printing Office, 1938.

Netzeband, W. F. *Method and Cost of Mining Zinc and Lead at Number One Mine, Tri-State Zinc and Lead District.* U.S. Bureau of Mines *Information Circular* No. 6113. Washington, Government Printing Office, 1929.

Oldright, G. L. *Zinc Smelting in the Horizontal Retort Fired with Natural Gas.* U.S. Bureau of Mines *Information Circular* No. 4336. Washington, Government Printing Office, 1948.

Perkins, Frances. *Conference on Health and Working Conditions in the Tri-State District, Joplin, Missouri, 1940.* Washington, U.S. Department of Labor, 1940.

Ruhl, Otto. *Tri-State Zinc-Lead Ore Reserves.* U.S. Bureau of Mines *Report of Investigations* No. 3922. Washington, Government Printing Office, 1946.

Ruhl, Otto, Simeon Allen, and Stephen P. Holt. *Zinc-Lead Ore Reserves of the Tri-State District, Missouri-Kansas-Oklahoma.* U.S. Bureau of Mines *Report of Investigations* No. 4490. Washington, Government Printing Office, 1949.

Sansom, Frank W. *Milling Practice at the Netts Mine of the Eagle Picher Lead Company at Picher, Oklahoma.* U.S. Bureau of Mines *Information Circular* No. 6342. Washington, Government Printing Office, 1930.

Santmeyers, Riegard M. *The Lead Industry.* Washington, U.S. Department of Commerce, 1925.

Sayers, R. R., et al. *Silicosis and Tuberculosis Among Miners of the Tri-State District of Oklahoma, Kansas, and Missouri.* U.S. Bureau of Mines *Technical Paper* No. 645. Washington, Government Printing Office, 1928.

Siebenthal, C. E. *Mineral Resources of Northeast Oklahoma.* U.S. Geological Survey *Bulletin* No. 340. Washington, Government Printing Office, 1908.

———. *Origin of the Zinc and Lead Deposits of the Joplin Region: Missouri, Kansas, and Oklahoma.* U.S. Geological Survey *Bulletin* No. 606. Washington, Government Printing Office, 1915.

Smith, W. S. Tangier. *Lead and Zinc Deposits of the Joplin District.* U.S. Geological Survey *Bulletin* No. 213. Washington, Government Printing Office, 1930.

———. *Water Resources of the Joplin District.* U.S. Geological Survey *Paper* No. 145. Washington, Government Printing Office, 1905.

———, and C. E. Siebenthal. *Geological Atlas of the United States, Joplin District Folio,* U.S. Geological Survey *Report* No. 148. Washington, Government Printing Office, 1914.

Thompson, J. W. *Abstracts of Current Decisions on Mines and Mining, December 1913 to September 1914.* Washington, U.S. Bureau of Mines, 1915.

Tri-State Zinc Company and International Union of Mine, Mill, and Smelter Workers, Affiliated with the C.I.O. Decisions and Orders of the National Labor Relations Board. XXXIX. Washington, Government Printing Office, 1942.

U.S. Bureau of Indian Affairs. *Lead and Zinc Mining Operations and Leases, Quapaw Agency: Regulations of the Indian Service.* Washington, Government Printing Office, 1934.

U.S. Bureau of Labor Statistics. *Employment and Payrolls, January, 1953.* Washington, Government Printing Office, 1953.

———. *Employment, Hours, and Earnings, Lead and Zinc Mining. Bulletin* No. 103. Washington, Government Printing Office, 1953.

———. *Hours and Earnings, Industry Report, February, 1953.* Washington, Government Printing Office, 1953.

———. *Productivity Trends, 1939–1950.* Washington, Government Printing Office, 1952.

———. *Wages and Hours of Labor in Metalliferous Mines, 1924. Bulletin* No. 394. Washington, Government Printing Office, 1925.

———. *Wages and Hours of Labor in Metalliferous Mines, 1924 and 1931. Bulletin* No. 573. Washington, Government Printing Office, 1932.

———. *Wages in Lead and Zinc Mines and Mills in the Tri-State Area, June, 1943. Bulletin* No. 1578. Washington, Government Printing Office, 1944.

———. *Wages in the Nonferrous Metals Industry, June, 1943. Bulletin* No. 765. Washington, Government Printing Office, 1944.

———. *Wage Structure of the Nonferrous Metals Industry, 1941–1942. Bulletin* No. 729. Washington, Government Printing Office, 1943.

———. *Wartime and Postwar Experiences of Nonferrous Metal Miners. Bulletin* No. 1887. Washington, Government Printing Office, 1945.

U.S. Bureau of Mines. *Crane and Chenoweth Mines, Jasper County, Missouri.* War Minerals *Report* No. 93. Washington, Government Printing Office, 1943.

———. *Little Mary Tailing Pile: Neck City District, Jasper County, Missouri.* War Mineral *Report* No. 86. Washington, Government Printing Office, 1943.

———. *Minerals Yearbook.* Washington, Government Printing Office, 1932–33.

———. *Oronogo-Webb City-Duenweg District of Jasper County, Missouri.* War Minerals *Report* No. 93. Washington, Government Printing Office, 1943.

———. *Oronogo-Webb City-Duenweg District of Jasper County, Missouri.* War Minerals *Report* No. 209. Washington, Government Printing Office, 1943.

——— and U.S. Geological Survey. *Mineral Resources of the United States.* Washington, Government Printing Office, 1883–1931.

U.S. Bureau of the Census. *Eighth Census of the United*

States: 1860, Population. I. Washington, Government Printing Office, 1863.
———. *Ninth Census of the United States: 1870, Population.* I. Washington, Government Printing Office, 1872.
———. *Tenth Census of the United States: 1880, Population.* I, II. Washington, Government Printing Office, 1882.
———. *Eleventh Census of the United States: 1890, Population.* I, II. Washington, Government Printing Office, 1891.
———. *Twelfth Census of the United States: 1900, Population.* I, II. Washington, Government Printing Office, 1901.
———. *Thirteenth Census of the United States: 1910, Population.* I, II. Washington, Government Printing Office, 1912.
———. *Fourteenth Census of the United States: 1920, Population.* I, II. Washington, Government Printing Office, 1921.
———. *Fifteenth Census of the United States: 1930, Population.* I, II. Washington, Government Printing Office, 1932.
———. *Sixteenth Census of the United States: 1940, Population.* I, II. Washington, Government Printing Office, 1941.
———. *Seventeenth Census of the United States: 1950, Population.* I, II. Washington, Government Printing Office, 1952.
———. *Special Report. Fifteenth Census of the United States: 1930, Unemployment,* I. Washington, Government Printing Office, 1931.
———. *Special Report. Religious Bodies, 1906.* I. Washington, Government Printing Office, 1910.
———. *Special Report. Religious Bodies, 1916.* I. Washington, Government Printing Office, 1919.
———. *Special Report. Religious Bodies, 1926.* I. Washington, Government Printing Office, 1930.
———. *Special Report. Religious Bodies, 1936.* I. Washington, Government Printing Office, 1941.
U.S. Congress. *Congressional Record.* 83 Cong., 2 sess., 1954, C.
U.S. Congress, House, *Current Conditions of the Lead and Zinc Mining Industry of the United States,* 83 Cong., 1 sess., *House Report No.* 688, 1953.
U.S. Congress, House, Hearings Before the Select Committee

on Small Business, *Problems in the Metal-Mining Industry (Lead, Zinc, and Other Metals)*, 83 Cong., 1 sess., 1953.

U.S. Congress, House, *War of Rebellion: Records of the Union and Confederate Armies*, Series I, Pt. 1, *Reports*, XXII, 50 Cong., 1 sess., *House Misc. Doc. No.* 585, 1888.

U.S. Congress, Senate, *Customs Tariff of 1846*, 62 Cong., 1 sess., *Sen. Doc. No. 71*, 1911.

U.S. Department of Agriculture. *Yearbook, 1941: Climate and Man.* Washington, Government Printing Office, 1941.

U.S. Statutes at Large. XXX, XLI, XLIX, LII, LXIII. Washington, Government Printing Office, 1889–1939.

U.S. Works Progress Administration. *Workers on Relief in the United States in March, 1935.* Washington, Government Printing Office, 1938.

Whitebird et al. v. *Eagle Picher Lead Co. Federal Reporter.* Second Series, XXVIII. St. Paul, West Publishing Co., 1929.

Wright, Clarence A. *Mining and Milling of Lead and Zinc Ores in the Missouri-Kansas-Oklahoma Zinc District.* U.S. Bureau of Mines *Bulletin* No. 154. Washington, Government Printing Office, 1918.

———. *Mining and Treatment of Lead and Zinc Ores in the Joplin District, Missouri, A Preliminary Report.* U.S. Bureau of Mines *Technical Paper* No. 41. Washington, Government Printing Office, 1913.

III. *State Documents*

Broadhead, Garland C. *Report of the Geological Survey of the State of Missouri, 1873–1874.* Jefferson City, Missouri Geological Survey, 1874.

———, F. B. Meek, and B. F. Shumard. *Report on the Geological Survey of the State of Missouri, 1855–1871.* Jefferson City, Missouri Geological Survey, 1873.

Buckley, E. R., and H. A. Beuhler. *The Geology of the Granby Area.* Jefferson City, Missouri Geological Survey, 1905.

Buehler, H. A. *Annual Report of the Missouri Geological Survey for 1907–1908.* Jefferson City, Missouri Geological Survey, 1908.

Constitution of the State of Oklahoma. Oklahoma City, n.p., 1948.

Dale, Phyllis, and J. O. Beach. *Mineral Production of Oklahoma, 1885–1949.* Oklahoma Geological Survey *Circular* No. 29. Norman, n.p., 1951.

Directory of the State of Oklahoma. Guthrie, Cooperative Publishing Co., 1951.

Forrester, J. D. *Mining and Mineral Resources of Missouri.* Technical Series *Bulletin* No. 73. Rolla, Missouri School of Mines and Metallurgy, 1948.

Forrester, J. D., and James F. Taylor. *A Comparative Analysis of Some Recent Mining Practices in the Tri-State Mining District.* Technical Series *Bulletin* No. 16. Rolla, Missouri School of Mines and Metallurgy, 1945.

Gould, Charles N., L. L. Hutchinson, and Gaylord Nelson. *Preliminary Report on the Mineral Resources of Oklahoma.* Oklahoma Geological Survey *Bulletin* No. 1. Norman, n.p., 1908.

Haworth, Erasmus. *Annual Bulletin on Mineral Resources of Kansas for 1897.* Lawrence, Kansas Geological Survey, 1898.

Haworth, Erasmus, W. R. Crane, and A. F. Rogers. *Special Report on Lead and Zinc.* Kansas Geological Survey *Report* No. 8. Lawrence, n.p., 1904.

Kansas Geological Survey. *Geophysical Investigations in the Tri-State Zinc and Lead Mining District.* Lawrence, n.p., 1942.

———. *Scenic Kansas. Bulletin* No. 36. Lawrence, n.p., 1935.

Kansas Mining Law, General Statutes of Kansas Annotated, 1949. Topeka, Voiland Printing Company, 1950.

Kansas State Commission of Revenue and Taxation. *Assessor's Manual, Kansas Tax Laws.* Topeka, Kansas State Printing Plant, 1950.

Lead and Zinc Mining Code, Oklahoma Statutes Annotated. St. Paul, West Publishing Co., 1939.

Mining Code, Revised Statutes of Missouri, 1899. Jefferson City, Tribune Printing Co., 1899.

Missouri Bureau of Labor Statistics. *Ninth Annual Report, Bureau of Labor Statistics, State of Missouri, 1887.* Jefferson City, Tribune Printing Co., 1887.

———. *Eleventh Annual Report, Bureau of Labor Statistics, State of Missouri, 1889.* Jefferson City, Tribune Printing Co., 1889.

Missouri Bureau of Mines. *Fifteenth Annual Report of the State Lead and Zinc Mine Inspector, State of Missouri, 1901.* Jefferson City, Tribune Printing Co., 1902.

———. *Seventeenth Annual Report of the State Lead and Zinc Mine Inspector, State of Missouri, 1903.* Jefferson City, Tribune Printing Co., 1904.

Missouri Planning Board. *A State Plan for Missouri: A Preliminary Report.* Jefferson City, n.p., 1934.

Missouri State Tax Commission. *Assessor's Manual, Missouri Tax Laws.* Jefferson City, Journal Printing Co., 1953.

Netzeband, W. F. *Technical Problems of the Tri-State Zinc-Lead Mining District.* Mineral Industries Conference *Proceedings.* Rolla, n.p., 1938.

Oklahoma Bureau of Mines. *First Annual Report of the Chief Inspector of Mines, State of Oklahoma, November 16, 1907, to October 13, 1908.* McAlester, n.p., 1908.

Oklahoma Geological Survey. *Mineral Production of Oklahoma from 1901. Bulletin* No. 15. Norman, n.p., 1912.

Redfield, John S. *Mineral Resources in Oklahoma.* Oklahoma Geological Survey *Bulletin* No. 42. Norman, n.p., 1927.

Revised Statutes of the State of Missouri, 1939. Jefferson City, Journal Printing Co., 1939.

Snider, Luther C. *Preliminary Report on the Lead and Zinc of Oklahoma.* Oklahoma Geological Survey *Bulletin* No. 9. Guthrie, Cooperative Publishing Co., 1912.

Steele, James H. *The Joplin Zinc District of Southwest Missouri.* Colorado School of Mines *Bulletin* No. 1. Boulder, n.p., 1900.

Swallow, George G. *The First and Second Annual Reports of the Geological Survey of Missouri.* Missouri Geological Survey. Jefferson City, James Lusk Co., 1855.

Williams, Charles P. *Industrial Report on Lead, Zinc, and Iron.* Jefferson City, Missouri Geological Survey, 1877.

Winslow, Arthur. *Lead and Zinc Deposits.* Jefferson City, Missouri Geological Survey, 1894.

IV. *Books*

Adams, James T., ed. *Dictionary of American History.* 6 vols. New York, Charles Scribners' Sons, 1942.

Allen, Henry J. *The Party of the Third Part: The Story of the Kansas Industrial Relations Court.* New York, Harper Co., 1921.

Allison, Nathaniel T. *History of Cherokee County, Kansas and Representative Citizens.* Chicago, Biographical Publishing Co., 1904.

Barnes, C. R., ed. *The Commonwealth of Missouri.* St. Louis, Bryan, Brand and Co., 1877.

Beatty, C. E., and J. F. Snow. *Joplin Missouri Mining and Industrial Interests.* Kansas City, Hudson-Kimberly Co., 1890.

Blackmar, Frank W., ed. *Kansas: A Cyclopedia of State History.* 2 vols. Chicago, Standard Publishing Co., 1912.

Bonnin, Gertrude C., et al. *Oklahoma's Poor Rich Indians.* Philadelphia, Office of Indian Rights Association, 1924.

Britton, Wiley. *Pioneer Life in Southwestern Missouri.* Kansas City, Smith-Grieves Co., 1929.

Conard, Howard L. *Encyclopedia of the History of Missouri.* 6 vols. New York, Southern History Company, 1901.

Connelley, William E. *A Standard History of Kansas and Kansans.* 4 vols. Chicago, Lewis Publishing Co., 1918.

———. *Quantrill and the Border Wars.* Cedar Rapids, Torch Press, 1910.

Dana, Daniel. *Journal of a Tour in the Indian Territory, 1844.* New York, Board of Missions of Methodist Episcopal Church, 1844.

Dans, Edward S. *A Textbook of Mineralogy.* New York, John Wiley and Sons, Inc., 1926.

Davidson, Lallah S. *South of Joplin: Story of a Tri-State Diggins'.* New York, W. W. Norton Co., 1939.

Dewitz, Paul W. *Notable Men of Indian Territory at the Beginning of the Twentieth Century.* Muskogee, Southwest Historical Co., 1905.

Draper, Mabel H. *Though Long the Trail.* New York, Rinehart and Co., 1946.

Draper, William R. and Mabel H. *Old Grubstake Days in Joplin.* Girard, Haldeman-Julius Co., 1946.

Emmons, William H., et al. *Geology: Principles and Processes.* New York, McGraw-Hill Book Co., 1939.

Federal Writers' Project. *Labor History of Oklahoma.* Oklahoma City, Works Progress Administration, 1939.

Finlay, James R. *The Cost of Mining.* New York, McGraw-Hill Book Co., 1909.

Foreman, Grant. *Indians and Pioneers.* Norman, University of Oklahoma Press, 1936.

Fowler, G. M., and J. P. Lyden. *The Ore Deposits of the Tri-State District.* New York, American Institute of Mining and Metallurgical Engineers Technical Press, 1932.

Garrison, Frank L. *Zinc and Lead Deposits of Southwestern Missouri.* San Francisco, Mining and Science Press, 1908.

Gent, Ernest. *A Review of the Zinc Industry in 1952.* New York, American Zinc Institute Press, 1953.

———. *The Zinc Industry: A Mine to Market Outline, 1949.* New York, American Zinc Institute Press, 1949.

———. *The Zinc Sheet: American Zinc Institute.* New York, American Zinc Institute Press, 1933.

———. *Thirty Years of Service: American Zinc Institute.* New York, American Zinc Institute Press, 1948.

Gist, Noel F., ed. *Missouri: Its Resources, People, and Institutions.* Columbia, University of Missouri Press, 1950.

Groneis, Carey, and William C. Krumbein. *Down to Earth, An Introduction to Geology.* Chicago, University of Chicago Press, 1936.

Gwynn, J. K. *Minerals and Mining.* St. Louis, Missouri Immigration Society, 1888.

Hamilton, Alice. *Industrial Poisons in the United States.* New York, MacMillan Co., 1925.

History of Newton, Lawrence, Barry, and McDonald Counties of Missouri. Chicago, Goodecoed Publishing Co., 1888.

Holibaugh, John R. *The Lead and Zinc Mining Industry of Southwest Missouri and Southeast Kansas.* New York, Scientific Publishing Co., 1895.

Hoover, Herbert, and L. H. Hoover, trans. *Georgius Agricola, De re Metallica.* London, n.p., 1912.

Horn, Orlando C. *Lead, the Precious Metal.* New York, Century Co., 1924.

Houck, Louis, ed. *The Spanish Regime in Missouri.* 2 vols. Chicago, Donnelley and Sons Co., 1909.

Hurlbut, S. L. *Tri-State Ore Formations.* Joplin, American Zinc, Lead, and Copper Journal Press, n.d.

Illustrated Historical Atlas of Jasper County Missouri. Joplin, Brink, McDonough and Co., 1876.

Ingalls, Walter R. *Lead and Zinc in the United States.* New York, Hill Publishing Co., 1908.

———. *World Survey of the Zinc Industry.* New York, McGraw-Hill Co., 1934.

The Klondike of Missouri. Kansas City, n.p., 1899.

Livingston, Joel T. *A History of Jasper County and Its People.* 2 vols. Chicago, Lewis Publishing Co., 1912.

Lloyd, E., and D. Baumann. *The Mineral Wealth of Southwest Missouri.* St. Louis, Democrat Printing Co., 1874.

Long, Stephen H. *The Stephen H. Long Expedition, 1819–1820.* Vol. XVII in Reuben Gold Thwaites, ed., *Early Western Travels, 1748–1846.* Cleveland, Arthur H. Clark Co., 1905.

McGregor, Malcolm G. *The Biographical Record of Jasper County, Missouri.* Chicago, Lewis Publishing Co., 1901.

McNeal, Tom A. *When Kansas Was Young.* New York, MacMillan Co., 1922.

Mathews, D. *Natural History of Jasper County.* Chicago, n.p. 1883.

Millis, Harry A., and Emily Clark Brown. *From the Wagner Act to Taft-Hartley.* Chicago, University of Chicago Press, 1950.

Mills, Lawrence. *Oklahoma Indian Land Laws.* New York, Thomas Law Book Co., 1924.

The Mineral Wealth of Southwest Missouri: Lead and Zinc Mines of Granby, Minersville, Joplin, Grove Creek, Stevens Mines, Thurman, Cornwell, Conley, and Others. Joplin, n.p., 1874.

Nicolson, C. W. *Recent Improvements in the Mining Practice of the Tri-State District.* New York, American Institute of Mining and Metallurgical Engineers Technical Press, 1938.

North, F. A. *The History of Jasper County, Missouri.* Des Moines, Mills and Co., 1883.

Parker, Nathan H. *Missouri as It is in 1867—An Illustrated Historical Gazetteer of Missouri.* Philadelphia, J. B. Lippincott Co., 1867.

———. *The Missouri Handbook: Information for Capitalists and Immigrants.* St. Louis, P. M. Pinckard Co., 1865.

Parsons, A. B., ed. *Seventy-five Years of Progress in the Mineral Industry, 1871–1946.* New York, American Institute of Mining and Metallurgical Engineers Technical Press, 1947.

Plat Book of Jasper County Missouri. Philadelphia, Northwest Publishing Co., 1895.

Pulsifer, William H. *Notes for a History of Lead and an Inquiry into the Development of the Manufacture of White Lead and Lead Oxide.* New York, D. Van Nostrand Co., 1888.

Rhyne, J. B. *Social and Community Problems of Oklahoma.* Guthrie, Cooperative Publishing Co., 1929.

Rickard, Thomas A. *A History of American Mining.* New York, McGraw-Hill Book Co., 1932.

Robinson, Edgar E. *The Presidential Vote, 1896–1932.* Stanford, Stanford University Press, 1934.

———. *They Voted for Roosevelt: The Presidential Vote, 1932–1944.* Stanford, Stanford University Press, 1947.

Robinson, Samuel. *A Catalog of American Minerals with Their Localities.* Boston, n.p., 1925.

Ross, Malcolm. *Death of a Yale Man.* New York, Farrar Publishing Co., 1939.

Ruhl, Otto. *The Rationale of Investment in Zinc Mining.* Joplin, Joplin News Press, 1910.

Ryan, Frederick L. *A History of Labor Legislation in Oklahoma.* Norman, University of Oklahoma Press, 1932.

———. *Problems of the Oklahoma Labor Market with Special Reference to Unemployment Compensation.* Oklahoma City, Semco Press, 1937.

Schoolcraft, Henry R. *A View of the Lead Mines in Missouri, Including Some Observations on the Mineralogy, Geology, Geography, Antiquities, Soil, Climate, Population, and Productions of Missouri.* New York, C. Wiley Co., 1819.

———. *Journal of a Tour into the Interior of Missouri and Arkansaw, from Potosi, or Mine a Burton, in Missouri Territory, in a Southwest Direction Toward the Rocky Mountains, Performed in the Years 1818 and 1819.* London, R. Phillips and Co., 1821.

———. *Scenes and Adventures in the Semi-Alpine Region of the Ozark Mountains of Missouri and Arkansas, Which were First Traversed by Desota in 1541.* Philadelphia, Lippincott Co., 1853.

Shaner, Dolph. *The Story of Joplin.* New York, Stratford House Inc., 1948.

Shepard, Charles U. *Reports Respecting Mineral Deposits in the States of Missouri and Illinois.* Boston, n.p., 1840.

Shinn, Charles H. *The Story of the Mine.* New York, D. Appleton Co., 1914.

Swallow, George C. *Geological Report of the Country Along the Line of the Southwestern Branch of the Pacific Railroad, State of Missouri.* St. Louis, George Knapp and Co., 1859.

Switzler, W. F. *Illustrated History of Missouri from 1554 to 1881.* St. Louis, C. R. Barnes Co., 1881.

Tri-State Survey Committee. *A Preliminary Report on Living, Working, and Health Conditions in the Tri-State Mining Area, Missouri, Oklahoma, and Kansas.* New York, Rotograph Co., 1939.

Tyler, Paul M. *From the Ground Up: Facts and Figures of the Mineral Industries of the United States.* New York, McGraw-Hill Book Co., 1948.

Waterhouse, S. *The Resources of Missouri.* St. Louis, Wiebusch Co., 1867.

Weidman, Samuel, C. F. Williams, and Carl O. Anderson. *Miami-Picher Zinc-Lead District.* Norman, University of Oklahoma Press, 1932.

Wilber, Charles D. *Mineral Wealth of Missouri.* St. Louis, E. J. Crandell Co., 1870.

Williams, Walter. *The State of Missouri.* Columbia, K. W. Stephens Press, 1904.

———, and Lloyd C. Shoemaker. *Missouri, Mother of the West.* 2 vols. New York, American History Society Inc., 1930.

Wilson, J. N., F. L. Clerc, and T. N. Davey. *Lead and Zinc Ores of Southwest Missouri Mines.* Carthage, Jasper County Democrat Press, 1887.

Witcombe, Wallace H. *All About Mining.* New York, Longmans Green Co., 1937.

V. *Periodicals*

Anderson, Carl O. "Flotation in the Tri-State District in 1925," *Bulletin of the American Zinc Institute,* Vol. VIII (May, 1926), 18.

———. "Growth of Flotation in the Tri-State District," *Engineering and Mining Journal,* Vol. XCI (June 12, 1926), 982.

———. "Recent Developments in Milling Practice in the Tri-State District," *Mining Congress Journal,* Vol. XV (November, 1929), 849–52.

———, and Elmer Isern. "Milling," *American Mining Congress and American Institute of Mining and Metallurgical Engineers Convention Booklet* (September 28, 1931), 22–24.

"Annual Average Metal Prices," *Engineering and Mining Journal,* Vol. CXLI (February, 1941), 55–56.

Bain, Foster. "The Origin of the Joplin Ore Deposits," *Engineering and Mining Journal,* Vol. LXXI (May 4, 1901), 557.

Bartlett, E. O., and P. L. Crossman. "Distribution of Lead and Zinc Ores Near Joplin, Missouri," *Engineering and Mining Journal,* Vol. LXVII (March 18, 1899), 321.

Bendelari, A. E. "Eagle-Picher, A Large Factor in the Lead Industry," *Mining Congress Journal,* Vol. XV (November, 1929), 857–59.

Bjorge, Guy N., and H. A. Walker. "Progress Reported in Methods and Equipment," *Mining and Metallurgy,* Vol. LXI (February, 1941), 66–68.

Boardman, D. F. "Power Systems of Mines of the Joplin District," *Engineering and Mining Journal*, Vol. LXXXVI (August 15, 1908), 327–29.

———. "Sheet-Ground Mine in Southwest Missouri," *Engineering and Mining Journal*, Vol. LXXXIV (November 8, 1907), 877–80.

Bourne, A. "Prairies and Barrens of the West," *American Journal of Science*, Vol. II (April, 1820), 30–35.

Bowman, R. G. "Lead Metallurgy," *Mining and Metallurgy*, Vol. LXI (February, 1941), 74.

Boyd, W. W. "The Joplin Mining District," *Canadian Mining Institute Transactions*, Vol. XV (1912), 617–30.

Brigham, G. W. "Lead and Zinc Mining in Oklahoma in 1910," *Engineering and Mining Journal*, Vol. XCI (January 7, 1911), 25.

Brittain, Doss. "Ground Breaking in the Joplin District," *Engineering and Mining Journal*, Vol. LXXXIV (August 10, 1907), 255–59.

———. "History of Smelting in the Joplin District," *Engineering and Mining Journal*, Vol. LXXXIV (November 9, 1907), 861–67.

———. "Milling Sheet Ground Ore in the Joplin District," *Engineering and Mining Journal*, Vol. LXXXIV (July 13, 1907), 59–64.

———. "Pumping Problems of the Joplin District," *Engineering and Mining Journal*, LXXXVI (August 1, 1908), 214–17.

———. "The Use of Natural Gas in the Joplin District," *Engineering and Mining Journal*, Vol. LXXXVI (September 19, 1908), 568.

Broadhead, Garland C. "Mines of Carterville, Jasper County," *Kansas City Review of Science*, Vol. VIII (1884), 70–77.

———. "Missouri Lead Smelters," *Engineering and Mining Journal*, Vol. XXV (February, 1883), 91.

———. "Notes on Surface Geology of Southwest Missouri and Southeast Kansas," *Kansas City Review of Science*, Vol. III (1879), 460.

———. "Old Granby Mines," *Engineering and Mining Journal*, Vol. XXXV (February, 1883), 406.

———. "Production of Zinc in Missouri," *American Geologist*, Vol. XII (1893), 274.

———. "Southwest Missouri Lead Interests," *Engineering and Mining Journal*, Vol. XXV (February, 1883), 73.

Bruce, James L. "Ore Dressing in the Joplin District," *Engineering and Mining Journal*, Vol. XCIII (February 24, 1912), 405–409.

Buckley, E. R. "Lead and Zinc Resources of Missouri," *Report of Proceedings of the American Mining Congress* (November 11, 1907), 38.

Burgess, Charles W. "Mining Costs in the Missouri-Kansas District," *Mining and Engineering World*, Vol. XXXVIII (April, 1913), 801–805.

Buskett, Evans W. "Sampling and Buying Ore in the Joplin District," *Engineering and Mining Journal*, Vol. LXXXVI (July 25, 1908), 190.

"By the Way," *Engineering and Mining Journal*, Vol. XCII (December 9, 1911), 1116, and Vol. CIII (June 9, 1917), 1939.

Campbell, John W., and E. C. Mabon. "Personnel and Safety," *Engineering and Mining Journal*, Vol. CXLIV (November, 1943), 118–22.

Caples, Russel B. "United States Zinc Survey," *Zn, Journal of the American Zinc Institute*, Vol. XXXI (April, 1953), 51.

Carter, Lane. "Economic Conditions in the Joplin District," *Engineering and Mining Journal*, Vol. XC (October 15, 1910), 759–61.

"Center Creek Mining Company," *Engineering and Mining Journal*, Vol. XLVII (March 30, 1889), 30.

Chapman, Temple. "Decline of Sheet Ore Mining at Joplin," *Engineering and Mining Journal*, Vol. XCI (May 6, 1911), 911.

———. "Departure in Sheet Ore Mining in the Joplin District." *Engineering and Mining Journal*, Vol. LXXXVII (May 8, 1909), 942.

———. "The Miami Zinc-Lead District," *Engineering and Mining Journal*, Vol. XCIII (June 9, 1912), 1146–47.

Chellson, H. C. "Dust Count Technique in Tri-State Mines," *Engineering and Mining Journal*, Vol. CXL (December, 1939), 29–33.

"Chitwood Hollow," *Engineering and Mining Journal*, Vol. LXIX (April 14, 1900), 535.

Clarke, S. S. "Mining Methods," *Engineering and Mining Journal*, Vol. CXLIV (November, 1943), 80–89.

———. "Surface Plant at a Typical Operation of the Eagle-Picher Lead Company," *Mining Congress Journal*, Vol. XV (November, 1929), 865–68.

Clerc, F. L. "Lewis Bartlett Process at Lone Elm," *Engineering and Mining Journal*, Vol. XL (July 4, 1885), 4.

Cobb, A. P. "The Importance of Zinc in Industry," *Mining Congress Journal*, Vol. XV (November, 1929), 840.

Coghill, W. H. "Milling in the Tri-State District," *Bulletin of the American Zinc Institute*, Vol. V (December, 1922), 86.

Coldren, Phil R. "Joplin-Miami District," *Engineering and Mining Journal*, Vol. CXIII (February 18, 1922), 307.

"Conditions of Zinc Mining at Aurora, Missouri," *Engineering and Mining Journal*, Vol. LXI (May 2, 1896), 422.

Conover, Julian D. "Zinc Mining in the Tri-State District," *American Zinc, Lead, and Copper Journal*, Vol. XX (June, 1928), 1–4.

Cooley, George T. "Dressing Zinc and Lead Ores in Southwest Missouri and Southeast Kansas," *Engineering and Mining Journal*, Vol. LVIII (July 7, 1894), 9.

"Cooperative Work: Tri-State Ore Producers' Association," *American Mining Congress and American Institute of Mining and Metallurgical Engineers Convention Booklet* (September 29, 1931), 26.

Crane, Clinton. "The Future of Lead and Zinc Markets," *Mining and Metallurgy*, Vol. LXI (October, 1940), 466–68.

Crane, W. R. "Methods of Prospecting and Mining in the Galena-Joplin District," *Engineering and Mining Journal*, Vol. LXXII (September 21, 1903), 360–62.

———. "Recent Changes in Mining and Milling in the Galena-Joplin Lead and Zinc District," *Engineering and Mining Journal*, Vol. LXXIV (September 27, 1902), 405–407.

———. "Slime Treatment in the Joplin Region," *Engineering and Mining Journal*, Vol. LXXVII (April 28, 1904), 683.

———, and G. I. Adams. "Lead and Zinc Mining in the Quapaw District," *Mines and Minerals*, Vol. X (May, 1907), 445–46.

"The Dayton Lead Mines," *Engineering and Mining Journal*, Vol. XXII (July 22, 1876), 61.

"Development and Underground Mining Practice in the Joplin District," *American Institute of Mining Engineering*, Vol. III (April, 1920), 48–50.

"Development of the Joplin District," *Engineering and Mining Journal*, Vol. LXVI (June 18, 1898), 725.

"Developments in the Joplin District," *Engineering and Mining Journal*, Vol. XCVI (October 11, 1913), 683–84.

"Editorial Correspondence," *Engineering and Mining Journal*, Vols. XCVI–CVIII (August 9, 1913–April 2, 1919).

"The Eight-Hour Law in Missouri," *Engineering and Mining Journal*, Vol. LXXX (December 2, 1905), 1026.

"Electrical Power at Joplin," *Engineering and Mining Journal*, Vol. LXXX (July 15, 1905), 64.

Ellis, E. E. "Occurrence of Commercial Zinc and Lead Ore Bodies in the Tri-State District," *Engineering and Mining Journal*, Vol. CXXI (January 30, 1926), 209–10.

Fellman, C. M. "Health and Safety in Mines," *Mining and Metallurgy*, Vol. LXII (February, 1941), 64–66.

Finlay, J. R. "Lead and Zinc Ores in Missouri," *Engineering and Mining Journal*, Vol. LXXXVI (September 26, 1908), 605.

Fowler, George M. "Geology of the Tri-State District," *Tulsa Geological Society Digest*. Vol. I (May, 1935), 43–47.

———, and Joseph P. Lyden. "The Miami-Picher Zinc-Lead District," *Economic Geology*, Vol. XXIX (June, 1934), 390.

———. "The Ore Deposits of the Tri-State District," *Economic Geology*, Vol. XXVIII (January, 1933), 75–81.

Garrison, Frank L. "Notes on Minerals, Including Joplin Zinc

Blendes," *Philadelphia Academy of Science Proceedings*, Vol. XIII (1907), 445–46.

———. "The Joplin Zinc District," *Mines and Minerals*, Vol. XX (1900), 462–63.

Gartung, F. H., G. M. Burke, and O. W. Bilharz. "Prospecting, Development, and Mining," *American Mining Congress and American Institute of Mining and Metallurgical Engineers Convention Booklet* (September 28, 1931), 18–21.

"Geology and the Mining Interests of Kansas," *Proceedings of the International Mining Congress*, Vol. IV (1901), 196–200.

Giessing, H. W. "Accident Prevention in the Tri-State District," *Mining Congress Journal*, Vol. XV (November, 1929), 853–56.

Gray, H. A., and R. J. Stroup. "Transportation," *Engineering and Mining Journal*, Vol. CXLIV (November, 1943), 90–92.

Guinn, J. B. "Missouri-Kansas Mining Methods and the Leasing System," *Engineering and Mining Journal*, Vol. LXVIII (August 5, 1899), 154.

Haley, J. Frank. "Churn Drilling in the Joplin District," *Engineering and Mining Journal*, Vol. LXXXIX (June 4, 1910), 1150.

Hallows, R. L. "Smelting," *Engineering and Mining Journal*, Vol. CXLIV (November, 1943), 108–15.

Hamilton, Alice. "A Mid-American Tragedy," *Survey Graphic*, Vol. XXIX (August, 1940), 432–37.

Hanley, H. R. "Economics and Metallurgy of Zinc," *Mining and Metallurgy*, Vol. LXII (February, 1941), 76–77.

Harbaugh, M. D. "Geology and Development of the Tri-State Zinc and Lead Mining District," *Tulsa Geological Society Digest*, Vol. I (May, 1935), 41–42.

———. "Geology and Ore Deposits," *American Mining Congress and American Institute of Mining and Metallurgical Engineers Convention Booklet* (September 28, 1931), 11–17.

Heap, R. R. "A Geological Drainage Problem: Miami Camp," *Engineering and Mining Journal*, Vol. XCVI (December 28, 1913), 1205–11.

———. "Zinc and Lead Mining in Oklahoma During 1913," *Engineering and Mining Journal,* Vol. XCVIII (January 10, 1914), 75–76.

Heckscher, A. "Zinc and Zinc Oxide in 1898," *Engineering and Mining Journal,* Vol. LXVIII (January 7, 1899), 20–21.

Hedburg, Eric. "The Missouri and Arkansas Zinc Mines at the Close of 1900," *American Institute of Mining and Engineering Transactions,* Vol. XXXI (1902), 379–404.

Heinz, C. E. "Yankee Ingenuity in a Tri-State Mill," *Engineering and Mining Journal,* Vol. CXXXVI (March 1, 1935), 135–38.

Henrich, Carl. "Zinc Blende Mines and Mining Near Webb City, Missouri," *Engineering and Mining Journal,* Vol. XXI (May 5, 1887), 3.

Herrick, R. L. "The Joplin Zinc District," *Mines and Minerals,* Vol. XXVIII (August, 1907), 145–57.

Hickman, Glenn A. "International Business Concern with Its Heart in the Magic Circle: Eagle Picher," *Mid-America Purchaser,* Vol. II (October, 1950), 1–4.

Higgins, Edwin. "Sheet Ground Mining in the Joplin District," *Mining and Engineering World,* Vol. XLIII (October, 1915), 523.

———. "Shoveling and Tramming in the Joplin District," *Engineering and Mining Journal,* Vol. C (October 23, 1915), 679.

"History of Eagle Picher," *Engineering and Mining Journal,* Vol. CXLIV (November, 1943), 68–79.

Hitchcock, C. K. "A Joplin Hand Jig," *Engineering and Mining Journal,* Vol. CI (March 11, 1916), 479–80.

Holibaugh, John R. "Early Mining in the Joplin District," *Engineering and Mining Journal,* Vol. LVIII (December 1, 1894), 508.

———. "Lead and Zinc Mining Industry of Missouri and Kansas in 1892," *Engineering and Mining Journal,* Vol. LV (February 18, 1893), 151.

———. "Zinc and Lead Mining in Missouri and Kansas in 1895," *Engineering and Mining Journal,* Vol. LX (July 27, 1895), 79.

———. "The Zinc Mining Industry of Southwest Missouri and Southeast Kansas," *Engineering and Mining Journal*, Vol. LVIII (October 27, 1894), 392.

Howell, S. P. "Prevention of Accidents with Explosives in the Tri-State Zinc and Lead Ore Producing District," *American Zinc, Lead, and Copper Journal*, Vol. XX (June, 1928), 4–6.

Hutchingson, Spenser. "Cost of Mining Zinc Ore in the Joplin District," *Engineering and Mining Journal*, Vol. LXXVII (February 25, 1904), 312–14.

Ihlseng, Axel O. "The Need of Improved Methods at Joplin," *Engineering and Mining Journal*, Vol. XC (July 16, 1910), 117.

Illidge, R. E. "Tri-State District Mining Methods," *Mining Congress Journal*, Vol. XV (November, 1929), 879–84.

"Imprisoned in a Joplin Mine," *Engineering and Mining Journal*, Vol. XCIII (August 5, 1911), 252.

"Inclined Shafts at Joplin," *Engineering and Mining Journal*, Vol. LXXXIX (January 22, 1910), 206.

Ingalls, Walter R. "Missouri and Kansas Zinc Production," *Engineering and Mining Journal*, Vol. LXXXIII (May 18, 1907), 939.

———. "Spelter Statistics for 1914," *Engineering and Mining Journal*, Vol. XCIX (March 6, 1915), 451–54.

———. "The Production of Zinc Ore in the United States," *Engineering and Mining Journal*, Vol. LXXIII (April 5, 1912), 476–78.

Isern, Elmer, "Central Milling," *Engineering and Mining Journal*, Vol. CXLIV (November, 1943), 93–105.

———. "Central Milling in the Tri-State District," *Engineering and Mining Journal*, Vol. CXXXI (January 26, 1931), 49–54.

"The Joplin and Girard Railroad," *Engineering and Mining Journal*, Vol. XXII (March 10, 1877), 328.

"Joplin Bucket Cars," *Engineering and Mining Journal*, Vol. XCIII (May 18, 1912), 979–80.

"The Joplin District," *Engineering and Mining Journal*, Vol. LXV (January 8, 1898), 35.

"Joplin District and the Coal Strike," *Engineering and Mining Journal*, Vol. LXXXI (April 28, 1906), 811.

"The Joplin District in 1903," *Engineering and Mining Journal*, Vol. LXXVII (January 7, 1904), 17.
"The Joplin District in 1906," *Engineering and Mining Journal*, Vol. LXXXI (February 3, 1906), 234–35.
"Joplin Method of Dumping Buckets," *Engineering and Mining Journal*, Vol. XCIV (August 3, 1912), 203.
"Joplin-Miami District," *Engineering and Mining Journal*, Vol. CIX (March 6, 1920), 629.
"The Joplin Mill Practice," *Engineering and Mining Journal*, Vol. LXXVIII (October 13, 1904), 579–80.
"Joplin Tram Car Concentrates," *Engineering and Mining Journal*, Vol. XCIII (April 20, 1912), 788.
"Joplin Type Bucket Elevators," *Engineering and Mining Journal*, Vol. XCIV (November 23, 1912), 977–78.
"Joplin Type Horse Whim," *Engineering and Mining Journal*, Vol. XCIV (August 3, 1912), 205.
"The Joplin White Lead Works," *Engineering and Mining Journal*, Vol. XXV (June 8, 1878), 394.
"The Joplin Zinc and Lead District in 1901," *Engineering and Mining Journal*, Vol. LXXIII (January 14, 1902), 29.
"The Joplin Zinc Ore Market," *Engineering and Mining Journal*, Vol. LXVII (February 4, 1899), 138.
"The Joplin Zinc Ore Market in 1899," *Engineering and Mining Journal*, Vol. LXIX (January 6, 1900), 18.
Just, Evan. "Tri-State District," *Engineering and Mining Journal*. CXLII (August, 1941), 147.
———. "Tri-State Strives to Maintain Output," *Engineering and Mining Journal*, Vol. CXLIII (December, 1942), 69–70.
"Kansas Mines and Minerals," *Transactions of the Kansas Academy of Science*, Vol. XVIII (1901), 200–207.
Keiser, H. D. "Retreating Coarse Tailings in the Tri-State District," *Engineering and Mining Journal*, Vol. CXXIII (April 9, 1927), 596–600.
Kitson, H. W. "The Mining Districts of Joplin and Southeast Missouri," *Engineering and Mining Journal*, Vol. CIV (December 22, 1917), 360, 1067–73, *Ibid.*, Vol. CV (March 2, 1918), 411–16; and *Ibid.*, Vol. CV (April 20, 1918), 727.

Koeker, K. L. "Has the Miami-Picher District Passed the Zenith?" *Engineering and Mining Journal*, Vol. CXVIII (April 15, 1924), 168–70.

Kurtz, Wade. "Mining Costs," *American Mining Congress and American Institute of Mining and Metallurgical Engineers Convention Booklet* (September 28, 1931), 24–25.

"Labor Troubles at Joplin," *Engineering and Mining Journal*, Vol. LXXX (December 16, 1905), 1121.

"Labor Troubles in Mining," *Engineering and Mining Journal*, Vol. C (July 10, 1915), 77–78.

Landis, H. K. "Zinc Mining in Missouri," *Engineering and Mining Journal*, Vol. LXI (January 11, 1896), 39.

Lanza, Anthony J. "Tuberculosis in the Joplin District," *Engineering and Mining Journal*, Vol. XCIX (February 13, 1915), 331–33.

"Lead," *Engineering and Mining Journal*, Vol. II (June, 1866), 342.

"Lead and Zinc Ores in Southwest Missouri," *Engineering and Mining Journal*, Vol. XLIII (June 4, 1887), 398.

"Lead Mining in Southwest Missouri," *Debow's Review*, Vol. XVIII (January, 1853), 389–91.

"Lead Poisoning in the Mining of Lead," *Mining Congress Journal*, Vol. XIII (March, 1927), 245.

Lippincott, Isaac. "Industrial Influence of Lead in Missouri," *Journal of Political Economy*, Vol. XX (July, 1912), 695–715.

McColgin, L. H. "Marketing the Concentrates," *American Mining Congress and American Institute of Mining and Metallurgical Engineers Convention Booklet* (September 28, 1931), 27.

McCormack, R. R. "Tri-State Mine Supply," *Engineering and Mining Journal*, Vol. CXLIV (November, 1943), 116–17.

McCuskey, E. W. "Tri-State Flood Control Cuts Mining Pumping," *Engineering and Mining Journal*, Vol. CXXXVI (July, 1935), 319.

MacKallor, D. C. "Mexican Operation," *Engineering and Mining Journal*, Vol. CXLIV (November, 1943), 106.

Marvin, Theodore. "Tri-State Mining Methods," *Explosives Engineer*, Vol. II (April, 1924), 109–14.
"Metallurgical Laboratory at Oklahoma School of Mines," *Engineering and Mining Journal*, Vol. XCII (July 15, 1911), 117.
"Miami District, Oklahoma," *Engineering and Mining Journal*, Vol. LXXXIX (January 8, 1910), 73.
"Mine Owners of Joplin," *Engineering and Mining Journal*, Vol. L (July 26, 1890), 96.
"The Mineral Resources of the Joplin District, Missouri," *Engineering and Mining Journal*, Vol. XLIX (March 1, 1890), 26.
"Mineral Wealth of Missouri," *Western Journal and Civilian*, Vol. VI (1853), 229–34.
"Mining Conditions in the Joplin District," *Engineering and Mining Journal*, Vol. LXXXIX (June 4, 1910), 1114.
"Mining News," *Engineering and Mining Journal*, Vols. XXIV–CXXXVI (July 21, 1877–July 1, 1935).
"Missouri Labor," *Engineering and Mining Journal*, Vol. LXXX (November 11, 1905), 902.
"The Missouri Lead and Zinc Company's Plant," *Engineering and Mining Journal*, Vol. LXIX (June 2, 1900), 648–49.
"Missouri Lead Mines," *Engineering and Mining Journal*, Vol. XXIII (April 21, 1877), 261.
Morford, A. L. "Zinc Mining at Galena, Kansas," *Engineering and Mining Journal*, Vol. XCIII (December 16, 1911), 1171.
Morton, Edward D. "Arizona Operation," *Engineering and Mining Journal*, Vol. CXLIV (November, 1943), 107.
Moseley, G. W. "Lead in Southwest Missouri, Moseley's Mines," *Western Journal and Civilian*, Vol. XIII (January, 1855), 119.
Moseley, William S. "Paper on the Lead Mines of the Southwest," *Western Journal and Civilian*, Vol. IV (1851), 412.
———. "Smelting Lead in Southwest Missouri," *Western Journal and Civilian*, Vol. III (1850), 412–13.
Naething, Foster S. "Ores of the Joplin Region," *Engineering and Mining Journal*, Vol. CXXIII (April, 1927), 575.
———. "The Oklahoma-Kansas-Missouri Zinc-Lead Field,"

Engineering and Mining Journal, Vol. CXXII (October 16, 1926), 604–608.

Netzeband, William F. "An Example of Mining Lead and Zinc Ore at Picher, Oklahoma," *Engineering and Mining Journal*, Vol. CXXVII (May 18, 1929), 792–97.

———. "An Example of Prospecting and Valuing a Lead-Zinc Deposit," *Engineering and Mining Journal*, Vol. CXXVII (June 8, 1929), 913–18.

———. "Profit from Mineral Waste: Tri-State Tailings Yield Commercial Products," *Engineering and Mining Journal*, Vol. CXXXVIII (May, 1937), 251–54.

———. "Prospecting with the Long Hold Drill in the Tri-State Zinc and Lead District," *American Institute of Mining Engineering Transactions*, Vol. LXXV (1927), 295.

———. "Relation of Fracture Zones to Ore Bodies in the Tri-State District," *Mining and Metallurgy*, Vol. IX (October, 1928), 446–47.

———. "Surface Transport of Ore and Tailings in the Tri-State District," *Engineering and Mining Journal*, Vol. CXXXVI (December, 1935), 601–605.

———. "Truck Transport Serves Tri-State Mining," *Engineering and Mining Journal*, Vol. CXLI (January, 1940), 34.

———. "Underground Deep Hold Prospecting at the Eagle Picher Mines," *American Institute of Mining Engineering Transactions*, Vol. LXXV (1927), 35–41.

———, and Howard O. Gray. "An Open-Pit Zinc-Lead Mine in the Tri-State District," *Engineering and Mining Journal*, Vol. CXL (June, 1939), 52–55.

"New Mineral Discoveries," *Western Quarterly*, Vol. I (July, 1868), 89–92.

"News From Washington," *Engineering and Mining Journal*, Vols. CXIII–CXXIII (March 18, 1922–May 14, 1927).

Nicholson, Frank. "Progress of the Zinc Industry in Missouri During 1903," *Engineering and Mining Journal*, Vol. LXXVII (January 14, 1904), 80.

———. "The Missouri-Kansas Zinc District," *Engineering and Mining Journal*, Vol. LXVII (June 10, 1899), 680.

"Notes on Zinc-Lead Mining in Missouri," *Engineering and Mining Journal*, Vol. XC (December 3, 1910), 1110.

"Oklahoma Mining Law," *Engineering and Mining Journal*, Vol. XCVII (March 28, 1914), 655.

"Oklahoma School of Mines," *Engineering and Mining Journal*, Vol. LXXXVIII (August 7, 1909), 278.

"Oklahoma's New Lead-Zinc District," *Engineering and Mining Journal*, Vol. LXXXVII (March 6, 1909), 496.

"Oklahoma's New School of Mines," *Engineering and Mining Journal*, Vol. CIX (February 28, 1920), 575.

"The Old Granby Mines, Newton County," *Engineering and Mining Journal*, Vol. XXV (June 15, 1878), 406.

"Opposition to Removing Quapaw Restrictions Develops," *Engineering and Mining Journal*, Vol. CIX (June 19, 1920), 1377.

"The Ore Market of the Joplin District in 1896," *Engineering and Mining Journal*, Vol. LXIII (January 2, 1897), 20–21.

"Origin of Mississippi Valley Lead and Zinc Deposits," *Engineering and Mining Journal*, Vol. CII (July 15, 1916), 150–51.

Palmer, Ernest J. "The Mines and Minerals of the Tri-State District," *Rocks and Minerals*, Vol. XIV (February, 1939), 35.

Pehrson, Elmer W. "Flow of Lead into World Trade in 1928," *Mining Congress Journal*, Vol. XV (November, 1929), 835–39.

Perkins, Edwin T. "Mining and Smelting at Granby, Missouri," *Engineering and Mining Journal*, Vol. LXXXIV (July 8, 1907), 388–90.

Perry, John. "Perry's Lead Mine," *Western Journal and Civilian*, Vol. I (1848), 608–13.

"Petroleum and Natural Gas in Kansas," *Engineering and Mining Journal*, Vol. LXXII (January 21, 1901), 397.

Pickerell, Albert G. "Lead Miners in Missouri," *New Republic*, Vol. XCII (August 11, 1937), 12–13.

Plumb, C. H. "The Joplin District Under the New Tariff Bill," *Engineering and Mining Journal*, Vol. LXXXVIII (December 25, 1909), 1285.

"Prevention of Lead Poisoning," *Engineering and Mining Journal*, Vol. XXVI (September 14, 1878), 184.

Pulsifer, H. B. "Development of Lead Smelting in Missouri," *Mining and Engineering World*, Vol. XL (June, 1914), 1148.

Raymond, Rossiter W. "American Lead Mines," *Engineering and Mining Journal*, Vol. X (February, 1871), 105.

———. "Missouri Lead Ores," *Engineering and Mining Journal*, XXVIII (August, 1879), 123.

———. "Notes on the Zinc Deposits of Southern Missouri," *American Institute of Mining Engineering Transactions*, Vol. VIII (May, 1879), 165–68.

———. "Notes on the Zinc Deposits of Southern Missouri," *Engineering and Mining Journal*, Vol. XXVIII (October, 1879), 240–41.

Rice, Claude T. "A Joplin Car for Boulders," *Engineering and Mining Journal*, Vol. XCIII (April 6, 1912), 690.

———. "Joplin Hoisting Bucket Hooks," *Engineering and Mining Journal*, Vol. XCIII (June 1, 1912), 1075.

———. "Joplin Jigging Practice," *Engineering and Mining Journal*, Vol. XCIII (May 18, 1912), 983–84.

Roosevelt, Ralph M. "The Use of Lead in Industry," *Mining Congress Journal*, Vol. XV (November, 1929), 841.

Rothwell, R. P. "The Lead and Zinc Ores of Southwest Missouri," *Engineering and Mining Journal*, Vol. XLIII (June, 1887), 397–98.

"Routine of Joplin Hoisting," *Engineering and Mining Journal*, Vol. XCIV (September 14, 1912), 491–92.

Ruhl, Otto. "Keeping Mining Costs at Joplin," *Engineering and Mining Journal*, Vol. XCII (December 9, 1911), 1133–36.

———. "Miami Lead and Zinc District in Oklahoma," *Engineering and Mining Journal*, Vol. LXXXVI (November 7, 1908), 910.

———. "Zinc in Missouri, Kansas, and Oklahoma," *Engineering and Mining Journal*, Vol. CLXVI (August 17, 1918), 302.

———, and O. W. Bilharz. "History of the Tri-State District," *American Mining Congress and American Institute of Mining and Metallurgical Engineers Convention Booklet* (September 28, 1931), 5–9.

Sansom, Frank W. "Core Drills in the Joplin District," *Engineering and Mining Journal*, Vol. XCI (April 1, 1911), 386.

———. "Milling Practice at the Eagle-Picher Lead Company," *Mining Congress Journal*, Vol. XV (November, 1929), 869.

Schaeffer, John A. "What Becomes of Lead Concentrates," *American Mining Congress and American Institute of Mining and Metallurgical Engineers Convention Booklet* (September 28, 1931), 33–37.

Snider, Luther C. "Oklahoma Lead and Zinc Fields," *Engineering and Mining Journal*, Vol. XCII (December 23, 1911), 1228–30.

Stone, Irene G. "The Lead and Zinc Field of Kansas," *Kansas Historical Collections*, Vol. VII (1902), 243–60.

Stroup, R. R. "Exploration Methods at the Eagle Picher Lead Company," *Mining Congress Journal*, Vol. XV (November, 1929), 860–64.

"Tailings from the Joplin District," *Engineering and Mining Journal*, Vol. LXXXIX (June 18, 1910), 1261.

Tarr, William A. "The Miami-Picher Zinc-Lead District," *Economic Geology*, Vol. XXVIII (August, 1933), 463–79.

———. "The Miami-Picher Zinc-Lead District," *Economic Geology*, XXIX (December, 1934), 779–80.

Tarver, Mark, and T. F. Risk. "Geological Survey of the State of Missouri," *Western Journal and Civilian*, Vol. III (1850), 76–83.

———. "Mineral Wealth of Missouri," *Western Journal and Civilian*, Vol. VI (1853), 229–34.

Terrell, A. D. "Smelting Practices," *American Mining Congress and American Institute of Mining and Metallurgical Engineers Convention Booklet* (September 28, 1931), 28–32.

Titcomb, Harold A. "Cost of Prospecting by Drilling in Joplin District," *Engineering and Mining Journal*, Vol. LXIX (April 7, 1900), 405.

———. "The Missouri-Kansas Zinc and Lead Mines," *Engineering and Mining Journal*, Vol. LXX (July 28, 1900), 98.

"The Tri-State Zinc-Lead District," *Kansas Geological Society Guidebook* (1937), 96–98.

"Tuberculosis in the Joplin District," *Engineering and Mining*, Vol. XCIX (February 13, 1915), 331.

Wampler, O. N. "Safety and Industrial Relations at Eagle Picher," *Mining Congress Journal*, Vol. XV (November, 1929), 876–78.

Waring, W. G. "The Zinc Ores of the Joplin District," *American Institute of Mining Engineering Transactions*, Vol. LVII (1917), 657–70.

———. "The Purchase of Zinc Ores in the Joplin District," *Engineering and Mining Journal*, Vol. LXX (July 14, 1900), 38–39.

"Washington Letter," *Engineering and Mining Journal*, Vol. CXXIII (May 14, 1927), 816.

Wells, T. L. "The Tri-State Today," *Explosives Engineer*, Vol. XIX (October, 1941), 306–11.

Wittlesey, Charles C. "Missouri and Its Resources," *Hunts Merchants Magazines*, Vol. VIII (1843), 535.

"Why Mining at Joplin Does Not Languish," *Engineering and Mining Journal*, Vol. XCVII (January 24, 1914), 232–33.

Williams, C. F. "The Leasing System," *American Mining Congress and American Institute of Mining and Metallurgical Engineers Convention Booklet* (September 28, 1931), 10.

———. "The Tri-State District from the Air," *Mining Congress Journal*, Vol. XV (November, 1929), 846–48.

Williams, C. P. "Some Points on Treatment of Lead Ores in Missouri," *Engineering and Mining Journal*, Vol. XXII (September 23, 1876), 220.

Williams, George K. "Mineral Production of Missouri in 1902," *Engineering and Mining Journal*, Vol. LXXVI (November 14, 1903), 739–40.

Wittich, Lucius. "Flotation in the Joplin District," *Engineering and Mining Journal*, Vol. CII (July 1, 1916), 45–46.

Wright, Clarence A. "Ore-Dressing in the Joplin District," *American Institute of Mining Engineering Transactions*, Vol. LVII (1917), 442–71.

Wright, Muriel. "Early Navigation and Commerce Along the

Arkansas and Red Rivers in Oklahoma," *Chronicles of Oklahoma*, Vol. VIII (March, 1930), 63–88.

Young, H. I. "Development and Underground Mining Practice in the Joplin District," *American Institute of Mining Engineering Transactions*, Vol. LVII (1917), 671–81.

———. "Mining Practice in the Joplin District," *Engineering and Mining Journal*, Vol. CIV (October 6, 1917), 595.

"Zinc Mining in the Joplin District," *Engineering and Mining Journal*, Vol. LXXXVI (August 22, 1908), 383.

"Zinc Smelting in Kansas," *Engineering and Mining Journal*, Vol. XC (October 15, 1910), 748.

"Zinc Smelting in Kansas and Oklahoma," *Engineering and Mining Journal*, Vol. XCIII (April 27, 1912), 849–51.

"Zinc Smelting in Oklahoma," *Engineering and Mining Journal*, Vol. LXXXII (August 25, 1906), 358.

Zook, Jesse A. "Conditions in the Joplin District," *Engineering and Mining Journal*, Vol. LXXIX (January 12, 1905), 86.

———. "Joplin District in 1905," *Engineering and Mining Journal*, Vol. LXXXI (January 6, 1906), 14–15.

———. "Joplin District in 1906," *Engineering and Mining Journal*, Vol. LXXXIII (January 5, 1907), 12–13.

———. "Joplin District in 1911," *Engineering and Mining Journal*, Vol. XCIII (January 6, 1912), 24–25.

———. "Joplin District in 1912," *Engineering and Mining Journal*, Vol. XCV (January 11, 1913), 22.

———. "Joplin District in 1913," *Engineering and Mining Journal*, Vol. XCVII (January 10, 1914), 74.

———. "Joplin District in 1914," *Engineering and Mining Journal*, Vol. XCIX (January 9, 1915), 65.

———. "Joplin in 1905," *Engineering and Mining Journal*, Vol. LXXXI (January 6, 1906), 15.

———. "Lead and Zinc in the Joplin District in 1916," *Engineering and Mining Journal*, Vol. CIII (January 6, 1917), 27.

———. "Lead and Zinc in the Joplin District in 1918," *Engineering and Mining Journal*, Vol. CVII (January 11, 1919), 59.

———. "Lead and Zinc in the Joplin District in 1919," *Engi-*

neering and Mining Journal, Vol. CIX (January 17, 1920), 383.
———. "The Missouri Ore Market," *Engineering and Mining Journal*, Vol. LXXIX (January 5, 1905), 16.
———. "Zinc and Lead in the Joplin District in 1908," *Engineering and Mining Journal*, Vol. LXXXVII (January 9, 1909), 69–72.
———. "Zinc and Lead in the Joplin District in 1909," *Engineering and Mining Journal*, Vol. LXXXIX (January 9, 1910), 72–73.

VI. *Newspapers*

Blackwell *Morning Tribune*. 1926–32.
Carl Junction *Plaindealer*. 1906–15.
Carl Junction *Socialist News*. 1906–1907.
Carthage *Evening Press*. 1886–99.
Carthage *Patriot*. 1876–78.
Carthage *Weekly Banner*. 1866–80.
Daily Oklahoman. 1897–1954.
Granby *Miner*. 1870–78.
Granby *News-Herald*. 1948–50.
Jasper County World. 1906–1907.
Joplin *Daily Herald*. 1875–1900.
Joplin *Globe*. 1897–1954.
Joplin *News Herald*. 1900–54.
Miami *Daily Record*. 1917–54.
Missouri Trades Unionist. 1911–21.
Neosho *Miner and Mechanic*. 1879–1950.
Neosho *Times*. 1869–1950.
Picher *King Jack*. 1922–35.
Picher *Tri-State Tribune*. 1935–53.
Webb City *Daily Register*. 1903–17.
Webb City *Daily Sentinel*. 1901–1906.

VII. *Unpublished Materials*

Edmondson, Ed. "Preliminary Report on Zinc in Oklahoma." Prepared by Representative Edmondson, Second District,

Oklahoma, for the Congress of the United States, House of Representatives, Washington, 1953. Mimeographed.

Friedman, Marion. "Annual Report of the Cherokee County Health Department." Prepared for the Cherokee County Health Department, Columbus, 1944. Mimeographed.

Gregory, Clay. "Report on the Herald Mine." M.A. thesis, Missouri School of Mines and Metallurgy, Rolla, Missouri, 1910.

Hall, Clifton. "Report of Tuberculin Testing in the Cherokee County Division of Tuberculosis Control." Prepared for the Kansas State Board of Health, Topeka, 1936.

"Health Status of the People of One Kansas County." Speech delivered before the Kansas Nutrition Conference, Topeka, 1941. Mimeographed.

Horner, C. K., and August F. Delaloge. "Investigation on Oklahoma Chats." B.A. thesis, Missouri School of Mines and Metallurgy, Rolla, Missouri, 1921.

Howes, Warren. "Salts in Tri-State Mill Waters." M.A. thesis, Missouri School of Mines and Metallurgy, Rolla, Missouri, 1930.

Jasper County Health Department. "Annual Report of the Jasper County Health Department." Webb City, 1940. Mimeographed.

Joplin Chamber of Commerce. "Joplin, Mineral Resources." Joplin, n.d. Mimeographed.

Joplin Chamber of Commerce. "Joplin, Missouri: History and Development." Joplin, 1933. Mimeographed.

Joplin Chamber of Commerce. "Mining History of Joplin." Joplin, 1950. Mimeographed.

Journal of A. L. Morford, retired journalist and writer on mining affairs, Galena, 1907–11.

"A Review of the Zinc Industry." Speech delivered before the American Zinc Institute in St. Louis by Evan Just, St. Louis. 1939. Mimeographed.

Long, Edgar C. "Reopening an Abandoned Tri-State Zinc Mine, Waco, Missouri." M.A. thesis, Missouri School of Mines and Metallurgy, Rolla, Missouri, 1939.

Mann, Horace T. "Losses in Smelting Missouri-Kansas Zinc

Ores." B.A. thesis, Missouri School of Mines and Metallurgy, Rolla, Missouri, 1910.
Netzeband, W. F. "The Role of Geology in Prospecting for Lead and Zinc in the Tri-State District." B.A. thesis, Missouri School of Mines and Metallurgy, Rolla, Missouri, 1927.
Oklahoma Health Department. "Tuberculosis in Oklahoma." Oklahoma City, 1945.
Oklahoma State Board of Health. "Oklahoma Health Bulletin No. Ten." Oklahoma City, 1952.
Stricklin, Cleo. "Tuberculosis in Cherokee County, Kansas." Prepared as a special report for the Kansas Tuberculosis and Health Association and the Public Health Department, Columbus, 1953.
Tri-State Ore Producers' Association. "Companies Operating in the Tri-State District of Oklahoma, Kansas, and Missouri." Picher, 1946. Mimeographed.
Tri-State Ore Producers' Association. "Tri-State District Lead and Zinc Mines Report." Picher, 1954. Mimeographed.
Tri-State Ore Producers' Association. "Why Conservation of Marginal Zinc and Lead Ores Means So Much to Our National Economy and Security." Picher, 1946.
Underwood, J. R. "Smelting of Lead Ores in Southern Missouri." B.A. thesis, Missouri School of Mines and Metallurgy, Rolla, Missouri, 1903.
Wright, Ira L. "Efficiency of Mill Practice in the Joplin District." B.A. thesis, Missouri School of Mines and Metallurgy, Rolla, Missouri, 1907.

VIII. *Interviews*

Interview with Leroy Adair, publisher of the Granby *News Herald*. Granby, Missouri, August 15, 1951.
Interview with Leo Nigh, superintendent at Eagle Picher West Side Mine. December 28, 1950.
Interview with H. A. Andrews, supervisor of leases on Quapaw Indian Lead and Zinc Lands. August 24, 1953.
Interview with L. C. Chamberlain, Eagle Picher mining engineer. Miami, Oklahoma, August 3, 1953.

Interview with Ira Davis, miner. Chitwood, Missouri, April 5, 1953.
Interview with Clarence Brown, driller and miner. Joplin, Missouri, April 2, 1953.
Interview with Lester Davis, miner and smelterman. Joplin, Missouri, August 5, 1953.
Interview with Roy Moore, miner and prospector. Purdy, Missouri, April 4, 1953.

IX. *Field Notes*

Field Notes. Galena, Kansas, December 30, 1950.
Field Notes. Joplin, Missouri, April 1, 1953.
Field Notes. Oronogo, Missouri, September 2, 1951.
Field Notes. Picher, Oklahoma, August 11, 1953.
Field Notes. Spurgeon, Missouri, August 25, 1951.
Field Notes. Treece, Kansas, December 29, 1950.

X. *Personal Correspondence*

Letter of Department of Education, State of Missouri, Jefferson City, October 8, 1953.
Letter of State Department of Education, State of Kansas, October 25, 1953.
Letter of State Board of Education, State of Oklahoma, Oklahoma City, October 15, 1953.
Letter of the Secretary of State, State of Kansas, Topeka, December 23, 1953.
Letter of the Secretary of State, State of Missouri, Jefferson City, January 4, 1954.

Index

Acme Mining Company: 170
Admiralty Zinc Company: 170
Air drills: 83, 92–93
Alba, Missouri: 12, 28
Allotments: 154
American Federation of Labor (AFL): 231, 233, 235, 237, 240
American Lead and Smelting Company: 172
American Lead and Zinc Company: 170
American Zinc Institute: 178
American Zinc, Lead and Smelting Company: 149, 167
Andrews, H. A.: 158
Andrews, H. D.: 38
Anglo-American Mining Syndicate: 166
Arkansas: 24
Artificial supports: 88
Asbury, Missouri: 28
Assay: 46
Assay content: 91
Atlantic and Pacific Railroad Company: 23, 117–18, 147, 150
Austin, Moses: 16, 20, 119

Badger camp: 28, 47, 82
Bain, H. F.: 5, 41, 79, 162
Bartlesville, Oklahoma: 8, 125
Bartlett, E. O.: 120
Bartlett process: 171
Baxter Springs camp: 80
Baxter Springs, Kansas: 13, 28, 30, 39, 49, 72, 81, 155, 178, 195, 235
Baxter Springs Mining Exchange: 155
Baumann, D.: 162
Bell Center, Missouri: 28
Belleville, Missouri: 28, 153
Benjamin, J. P.: 26
Bethel-Domando-Croesus Company: 170
Bilharz Mining Company: 172
Bird Dog Concentrator: 112
Bituminous coal: 122
Black blasting powder: 71
Black Eagle Mining Company: 233
Blacksmiths: 92
Blanket deposit: 106
Blast furnace: 23, 116
Blasting: 76, 83
Blende City, Missouri: 28, 37
Blow and Kennett Mining Company: 22–23, 117
Blue Card Union: 234, 235, 237, 239, 241
Blytheville, Missouri: 18, 28, 30, 32
Boston-Duenweg Company: 150
Boulder popping: 89
Bradbury, Alonzo: 32
Breast-Stope Method: 76
Brecciated and boulder ground: 10
Britton, Wiley: 7, 25, 127, 201

Broadhead, Garland C.: 77, 101, 104
Broadhead's *Report for 1873*: 39
Brown, Clarence: 50
Brown, Tom: 231
Bruce, James: 198
Bruno man: 197
Buck hammer: 102
Buck rocks: 264
Burch City, Missouri: 28
Burgess, Charles W.: 201

Cable-drawn ore cars: 84
Cameron Pump: 74
Campbell, David: 19–20, 114
Candles: 92
Caney, Kansas: 125
Cans: 88
Carbide mining lamps: 92
Cardin, Oklahoma: 7, 28, 40, 112, 172
Carl Junction, Missouri: 28, 37
Carl Junction *Socialist News*: 230
Carney Diggings: 28, 147, 165
Carpenter's rig: 42
Carter, Lane: 163, 202, 252
Carterville, Missouri: 28, 30, 37, 83, 167, 184
Carthage, Missouri: 28, 30, 32, 83, 160, 258
Case and Searge Lead Company: 121, 166
Cave-in: 90, 179, 180
Cave Springs, Missouri: 28
Caving: 88
Cedar Creek Diggings: 147
Center Creek camp: 3, 18, 22, 26, 30, 117
Center Creek Company: 175
Center Creek Mining Company: 165
Center Creek Mining Land Company: 148
Central and custom milling: 104, 111
Central City, Missouri: 28, 250
Central company mill: 130
Centralized milling: 68
Central mill: 96, 105
Central milling: 105, 111, 158
Century camp: 82
Chanute, Kansas: 125
Chapman and Riggins Land and Smelting Company: 147
Chapman, Temple: 183
Charcoal: 122
Cherokee County Health Department: 194
Cherokee County, Kansas: 3, 128, 203, 259, 272
Cherryvale, Kansas: 125
Chetopa, Kansas: 3
Childers Mining Company: 170
Childress, Harold: 172, 178
Chitwood, Missouri: 7, 28, 37, 96, 167
Chitwood camp: 82, 85
Chitwood Land and Mining Company: 85
Churn drill: 42, 46, 67, 196
Churn-drill prospecting: 84, 91
Churn-type drill: 45
Clary, Joseph: 180
Club Theater, Joplin: 38
Coal: 81, 125
Coal miners: 201
Cokey: 197, 200
Collinsville, Oklahoma: 126
Columbus, Kansas: 195
Commerce (Hattonville), Oklahoma: 28, 39, 83, 96, 158, 247, 267
Commerce field: 39, 172
Commerce Mining Company: 170, 172
Commerce Mining and Royalty Company: 48, 112
Commingling: 157, 158
Committee for Industrial Organization (CIO): 239, 241, 245
Commonwealth Mining Company: 170

Company of St. Philip: 15
Company of the West: 14
Company town: 255
Company union: 233
Concentrates: 91, 108, 113, 173, 197
Concentrating process: 109
Confederate Army: 25
Consolidated Lead and Zinc Company: 172
Consolidated Mining Company: 147
Consolidation: 68
Consultant facilities: 97
Co-operative water company: 85
Cordwood: 122
Cornish Pump: 74, 86
Corporate financing: 175
Cotton-rock roofs: 88
Cowskin (Elk) River: 3, 24
Cox Diggings: 28, 165
Cox, John C.: 18, 20, 31, 32
Cox mines: 20
Crawford County, Kansas: 201, 203
Crawford, Samuel: 155
Crestline, Kansas: 28
Cribbing: 73
Crozat, Anthony: 14, 16
Crude ores: 100
Crusher: 70
Cull man: 197
Cunningham, Leonidas P.: 167
Cupola method: 121
Custom milling: 111

Davidson, Lallah: 185, 239, 250
Davis, Cal: 247
Dawson, George: 270
Dayton camp: 28
Death of a Yale Man: 251
Deep shafting: 72
Denver Mining and Flotation Company: 110
Department of the Interior: 128, 156, 157
Department of Labor conference on working and housing conditions: 257
Depression: 174, 199, 263
De Re Metallica: 182
Diamond: 28
Diamond drill: 47
Diamond, Missouri: 161
District development since 1865: 27
District development to 1865: 14
Division of labor: 197
Douthat, Oklahoma: 28
Douthat camp: 82
Draper, Mabel: 258
Drift dimensions: 75
Drift floor: 95
Drifting: 67, 75, 93, 96, 179
Drifts: 84
Driller: 92, 98
Drill holes: 78
Drilling: 10, 44, 87
Drills: 93, 196
Drummond furnace: 115
Duenweg, Missouri: 28, 150
Dust control: 182, 189
Dynamite: 44, 83, 89, 181, 196

Eagle Picher Company: 95, 96, 112, 126, 151, 157, 171, 192, 229, 235, 241, 245, 253
Eagle Picher Lead Company: 166, 170
Eagle Picher Mining and Smelting Company: 92, 126
Eagle Picher West Side Mine: 95
Eagle Picher's central mill: 158, 246, 247, 267
East Joplin Mining and Smelting Company: 148
East St. Louis: 173
Edmondson, Ed: 269
Educational facilities: 258
Eight-hour law: 199
Elections: 257
Electric cap lamps: 92

Electric hoist: 98
Electricity: 83
Electric locomotive: 94
Electric railway: 82
Electrolytic process: 126
Elk (Cowskin) River: 3, 18
Empire City: 28, 37
Empire District Electric Power Company: 81
Emma Gordon Company: 88
Empire Mining and Smelting Company: 104, 165
Engineering and Mining Journal: 162
Engineers: 165
English mining syndicate: 165
English smelters: 175
Evans, F. W. "Mike": 233, 234, 236, 240
Ewert, Paul: 156
Explosives: 181

Fair Labor Standards Act: 200
Fall, Albert: 157
Farming: 198
Federal Labor Union: 229
Federal Mining and Smelting Company: 150, 170
Federated Mining Company: 203
Fidelity, Missouri: 28
Finlay, James R.: 105
Five Mile Creek: 3
Flatboating: 81
Flotation: 109
Food processing: 273
Foreign capital: 148, 165
Foreign imports: 174, 178
Foreign outlets: 174
Forrester, J. D.: 92
Fowler, George M.: 6
French Point, Missouri: 30, 117
Frisco Railroad: 96
Frontier V furnace: 114
Frye, J. H.: 161
Galena, Kansas: 28, 29, 30, 37, 39, 47, 81, 83, 90, 98, 99, 104, 111, 125, 126, 148, 165, 166, 178, 183, 192, 195, 231, 235, 251, 257, 258, 272
Galena camp: 11, 80, 128, 129
Galena lead smelter: 171
Gallagher, James: 165
Galvanizing: 123, 174
Gambling dens: 251
Geology: 27, 127
Geophone: 93
Geophysical explorations: 48
Georgia City, Missouri: 28
Giant powder: 71
Goldenrod Mining and Smelting Company: 170
Gougers: 76, 97
Gouging: 263
Gould, Jay: 164
Granby, Missouri: 21, 22, 25, 28, 30, 36, 111, 117, 121, 123, 150, 160, 165, 166, 204, 207, 251, 254
Granby *Miner*: 251
Granby Mining and Smelting Company: 48, 103, 105, 111, 120, 121, 124, 126, 130, 147, 150, 165, 166, 207
Grand Falls camp: 28, 80, 122, 166
Grand Falls power plant: 81
Grand River: 24
Great Circle mine: 12
Greene County, Missouri: 231
Grizzly: 102, 107
Groundwater: 95
Grubb Pump: 74
Grubstake: 161, 196, 198, 208
Guinn, J. B.: 153

Hamilton, Alice: 190
Hand drill: 71, 83
Hand jig: 102, 264
Hand labor: 93
Hand loading: 94
Hand pick: 101
Hand windlass: 67, 130, 264

Hannibal Lead Works: 207
Harklerode Company: 24, 117
Hartley-Grantham mine: 49
Hattonville camp: 39
Havre, France: 175
Haworth, Erasmus: 9, 11, 75, 168
Hazards: 179
Heading: 87
Health and sanitation programs: 97
Health and working conditions: 187, 189
Health services: 177
Hearth furnace: 118
Hecksher, A.: 202
Henrietta, Oklahoma: 126, 171
Hettig Company: 170
Hickman, Glenn: 233, 234, 242
Hill, William: 270
Hockerville, Oklahoma: 28
Hoister: 71
Hoisterman: 70, 77, 90, 92, 98, 197
Hoisting equipment: 67, 84, 161
Hoists: 39, 70
Holden Tract: 153
Holibaugh, John R.: 17
Hoover, Herbert: 182
Hopper: 107
Horse-drawn ore trucks: 82, 124
Horse hoister: 74, 130, 264
Horse whim: 70
Housing: 189, 254
Howe, John W.: 205
Hulsman, Ruth: 193
Hunt, Commonwealth, Kelton, and White Bird Mining Companies: 157
Hydroelectric plant: 80

Illustrated Historical Atlas of Jasper County: 19
Inclined shaft: 84
Indian Creek: 18
Indian lands: 127
Indian Rights Association: 157
Indians: 113, 153, 154
Indian Territory: 24, 39
Industrial poisoning: 190
Industrial Workers of the World (IWW): 227, 230
Ingalls, Walter R.: 164
International Union of Mine, Mill, and Smelter Workers (CIO): 229, 231, 232, 239, 240, 243, 244
Investment: 148, 169
Investors: 166, 169
Iola, Kansas: 125
Italian laborers: 89, 203

Jackson, Charles F.: 49
Jackson Diggings: 28
Jamison, Ray: 236
Jasper County, Missouri: 3, 20, 31, 38, 39, 116, 117, 147, 192, 204, 230, 259, 272
Jasper County Anti-Tuberculosis Society: 183
Jasper County Court: 35
Jasper County Socialist party: 227, 230
Jenkins Creek: 3
Jennings, Edmund: 17
Jig: 70
Jigging: 102, 108
Jones Creek: 3
Joplin, Harris: 19, 29, 32
Joplin, Missouri: 28, 30, 32, 34, 36, 37, 39, 42, 74, 81, 83, 96, 97, 99, 102, 103, 110, 111, 116, 121, 122, 125, 126, 128, 147, 148, 150, 160, 165, 166, 173, 178, 189, 192, 202, 204, 227, 231, 235, 251, 257, 258, 273
Joplin Business Club: 164
Joplin camp: 84
Joplin Colic: 38, 249, 254
Joplin County: 272
Joplin Creek: 3, 20, 114
Joplin Creek field: 31
Joplin Creek valley: 20, 31

Joplin district: 4, 34, 114, 120, 121, 128, 184
Joplin field: 147
Joplin-Girard Railroad: 81, 167
Joplin *Globe*: 231, 251
Joplin Lead Company: 122
Joplin Mill: 112
Joplin Mill Practice: 106
Joplin mines: 124
Joplin Mining Stock Exchange: 165
Joplin *News Herald*: 231, 251
Joplin Ore Producers' Association: 176
Joplin Thousand Acre Tract: 37, 150, 255
Joplin Trades Assembly: 228
Joplin White Lead: 121
Jumboes: 126
Jumbo Mobile Drill: 93
Just, Evan: 244

Kansas: 82, 125, 128, 179, 199, 203, 209
Kansas City, Missouri: 227
Kansas City bottoms: 20, 69, 147
Kansas City, Fort Scott and Gulf Railroad: 81, 82
Kansas City, Fort Scott and Memphis Railroad: 162
Kansas City Mining Exchange: 148
Kansas coal fields: 122
Kansas Explorations Company: 154, 170
Kansas State Board of Health: 195
Kansas zinc smelters: 124
Keener, Oliver W.: 49
Kennett and Blow: 24
Keystone rig: 43
Klondike camp: 28, 47, 82
Klondike of Missouri: 38, 162, 169
Knights of Labor: 227

Labor: 199, 202, 203, 235
Labor force: 198
Labor movement: 209, 210, 231
Labor organizations: 205
Labor relations: 158
Labor unions: 196
Lackawanna Mining Company: 167
Laclede Lead and Zinc Company: 170
LaMotte, M.: 15
Land and royalty companies: 129, 160, 165–66, 173, 207
Landowners: 68
Land tenure: 127
Land-use pattern: 154
Lanza, Dr. Anthony J.: 184, 185, 187, 191, 250, 256
Lard-oil lamps: 92
Law, John: 14
Lawton camp: 28
Lead colic: 191
Lead fume: 120
Lead poisoning: 191
Lead refineries: 121
Lead smelting: 191
Leadville Diggings: 165
Leadville Hollow camp: 28, 30, 110, 116, 168
Leadville mines: 20, 28
Lead and zinc code: 200
Lead and zinc smelting: 126
Leaseholders: 111
Leasing: 48, 127
Lee, Levi: 19
Lehigh camp: 28, 90
Lewis-Bartlett process: 122–23
Lewis, G. T.: 120
Lewis, John L.: 239
Liberty, Kansas: 81
Lincolnville, Oklahoma: 8, 28, 39, 82
Liverpool, England: 175
Livingston, Joel T.: 18, 21, 31, 33
Livingston, Thomas: 21, 117
Lloyd, E.: 162
Lobbying: 178

Log furnace: 113, 114, 117
Lone Elm camp: 28, 37, 105
Lone Elm Smelting Works: 122
Long, Stephen H.: 17
Lowell, Kansas: 80
Lowell power plant: 81
Low-grade ores: 84
Lyden, Joseph P.: 6

Mabon, E. C.: 245
Machines: 93
McKee, Andrew: 21
Magmatic theory: 4, 6
Magnetic survey: 48
Manpower output: 92
Marginal ores: 112, 267
Market prices: 72, 100
Markets: 24, 100, 160, 173, 174
Mary M. Mining Company: 172
Mears, W. S.: 167
Mechanical drill: 83
Mechanical loading: 89, 92, 94
Mechanization: 194, 199
Melrose camp: 3
Memphis Railroad: 82
Metallurgists: 105, 120
Metallurgy: 123
Metropolitan Life Insurance Company: 177, 187
Miami, Oklahoma: 3, 28, 38, 83, 155, 158, 177, 258, 273
Miami camp: 39, 44, 48, 88
Miami *Daily Record*: 247
Miami field: 46
Migration: 261, 272
Mineral leasing: 127, 148
Mineral Wealth of Southwest Missouri: 162
Mill operator: 76
Millman: 98
Milling: 27, 41, 100–101, 272
Mills: 39
Mine accidents: 181
Mine accounting: 92
Mine fatalities: 179
Mine inspector: 179, 182
Mine inspector for Missouri: 149
Mine LaMotte: 16, 67
Mine pump: 74
Mine unionization: 229
Mine water: 84
Miners: 80, 199, 227, 260
Miners' Benevolent Association: 207
Miners' code: 206
Miner's consumption ("con"): 72, 93, 182
Miners' families: 183
Miners' housing: 186
Minersville, Missouri: 21, 22
Mining: 27, 41, 272; improved methods of, 97
Mining camps: 150
Mining codes: 127, 128
Mining engineers: 79, 151
Mining exchanges: 178
Mining land and royalty companies: 48, 68, 103, 104, 108, 127, 147, 148
Mining land system: 128
Mining law: 186
Mining leases: 103, 110
Mining operations: 25, 161
Mining society: 249
Mining techniques: 128
Mining tracts: 129
Mining World: 162
Missouri: 24, 82, 179
Missouri Bureau of Mines and Geology: 48
Missouri, Kansas and Texas Railroad: 82
Missouri-Kansas Zinc Miners' Association: 175
Missouri labor commissioner: 208, 251, 261
Missouri Lead and Zinc Company: 150, 151, 255
Missouri Legislature: 199
Missouri and Northwest Railroad: 81
Missouri Pacific Railroad: 82

Missouri Trade Unionist: 229, 230
Mobile caterpillar slusher: 94
Mobile churn drill: 42
Moffet, A. E.: 122
Moffett, E. R.: 31, 35
Moffett and Sergeant Mining Company: 148
Moffett-Sergeant lease: 32
Monkey Hill tract: 165
Montreal Company: 170
Moon Hill Diggings: 148
Moon Range Diggings: 28, 37
Moore, Roy: 50
Morford, A. L.: 155
Moseley, J. W.: 23, 24, 38
Moseley, William: 115
Moseley Lead Manufacturing Company: 117
Moseley's camp: 25
Mucker: 197
Mules: 74, 89, 94, 98
Muleskinner: 197
Murphy, Patrick: 32, 35, 165
Murphysburg: 28, 32, 33, 34, 253
Murphysburg Town Company: 32
Mutual Mining Company: 172

National Industrial Recovery Act: 200, 231
National Labor Relations Act: 240
National Labor Relations Board: 233, 234, 235, 240, 241, 242
National Zinc Company: 8
Natural gas: 81, 122
Natural pillars: 88
Neck City, Missouri: 3, 166
Negroes: 202
Nellie B. Mining Company: 172, 245
Neodesha, Kansas: 125
Neosho, Missouri: 23, 25, 28, 30, 115, 204, 258
Neosho River: 3
Netta mine: 106
Newton County, Missouri: 3, 19, 22, 24, 31, 36, 38, 115, 117, 120, 147, 204
Newtonia, Missouri: 3
New York and St. Louis Mining and Manufacturing Company: 150
Nicolson, Frank: 162
"Nigger Diggins": 20
Nolan, Joe: 235, 237
Norman, Kelsey: 237, 240
North, F. A.: 21, 33
North American Exploration Company: 48
North Fork River: 18
North Star Company: 153

O'Jack Mining Company: 97
Oklahoma: 38, 82, 83, 128, 179, 209
Oklahoma camps: 38
Oklahoma constitution: 199
Oklahoma County, Oklahoma: 231
Oklahoma field: 106, 111
Oliver's Prairie Mines: 118
Open-pit mining: 90
Operator association: 98
Orchard Diggings: 148
Ore buyers: 82, 173, 176
Ore cans: 77
Ore hauling: 82
Ore hopper: 90
Ore Producers' Association: 162
Ore trucks: 95
Organized labor: 247
Oronogo (Minersville), Missouri: 12, 17, 20, 21, 37, 81, 111, 113, 147
Oronogo Circle Mine: 96, 172
Osage River: 24
Ottawa County, Oklahoma: 3, 29, 38, 39, 154, 171, 192, 194, 201, 231, 235, 259, 272
Ozark hill country: 201, 255
Ozark Power and Water Company: 80

Ozark Uplift: 3, 4, 6

Parr Hill camp: 28
Parsons, Kansas: 82
Peacock Valley camp: 28, 90
Peoria, Oklahoma: 28, 39
Peoria camp: 8
Peoria Mining Land Company: 39
Perkins, Frances: 189, 204, 243
Phelps Mining Company: 97
Phoenix Lead and Zinc Mining and Smelting Company: 150
Picher, Oklahoma: 28, 40, 83, 88, 99, 106, 151, 154, 177, 183, 187, 195, 203, 229, 231, 235, 253, 255, 257, 272
Picher Clinic: 189, 190, 193
Picher Company: 208
Picher field: 28, 172
Picher *King Jack*: 251
Picher Lead Company: 40, 121, 122, 147, 148, 166, 207
Picher Mining Land and Smelting Company: 171
"Pick-handle Brigade": 235
Pierce City, Missouri: 28
Pig tail: 197
Pinkard Diggings: 28, 165
Pit-prospect method: 67
Pittsburg, Kansas: 6, 80, 81
Playter, Frank: 165
Playter Mining Company: 111
Pockety deposit fields: 80
"Poor Man's Camp": 68, 99, 103, 128, 147, 160, 206, 259
"Poor Man's District": 152
Population: 261, 272
Portable drill: 43
Porto Rico, Missouri: 28, 82
Powder: 42, 71
Powderman: 76, 92, 98
Powder monkey: 198
Power shovels: 94
Prairie Diggings: 22
Profits: 169
Prospect drilling: 40
Prospecting: 41, 47, 75, 93, 147, 196, 198
Prospecting records: 92
Prospectors: 46, 98, 252
Prosperity, Missouri: 28
Pull drifts: 46
Pump: 70, 196
Pumping: 67, 95, 161
Pumping machinery: 85
Pumping station: 85
Pyramiding of corporate control: 160
Pyramiding of mineral land leases and royalties: 149

Quantrill, Charles: 30
Quapaw, John: 153, 154
Quapaw, Oklahoma: 28, 38, 233
Quapaw Indian lands: 29, 154, 178
Quapaw mining land: 156
Quapaw Reservation: 128
Quapaws: 38, 152, 154, 155, 156, 157

Racine, Missouri: 28, 209
Railroads: 81, 96, 171
Ramage Company: 170
Rationale of Investment in Zinc Mining, The: 162
Reed, James: 244
Refinery: 116, 125
Refining process: 23, 113
Renault, Philip: 14–15
Reports of the Geological Survey of Missouri: 22
Research laboratory: 171
Respiratory diseases: 182
Retorts: 124
Reverberatory furnace: 116
Rex Mining and Smelting Company: 150
Rex Mining Company: 255
Rhyne, J. B.: 203
Richards Mining Company: 147

Riggins-Chapman Smelting Company: 103
Riley, John: 205
Riot: 147
Riverton, Kansas: 80, 195
Riverton power plant: 81
Rock dust: 93
Rock wool insulation plant: 171
Rockefeller, John D.: 164
Rogers, A. F.: 6
Roof-caving tests: 93
Roof trimmers: 180, 197
Ross, Malcolm: 251, 264
Rougher jig: 107
Roustabout: 197
Royalties: 68, 111, 127, 129, 151, 153, 158, 161, 168, 169, 181, 207
Rubberneck Tract: 166
Ruhl, Otto: 162, 168, 200, 203, 228
Ryan, Frederick L.: 185, 191, 227

Safety: 97
Safety programs: 93, 177, 179, 182
Saginaw, Missouri: 28
St. Francis River: 15
St. Louis, Missouri: 15, 24, 28, 81, 147, 165
St. Louis camp: 82
Saloons: 251
S. B. Corn Mining Company: 147
Schoolcraft, Henry R.: 15, 113–14
Scotch Eye Furnace: 119
Scotch hearths: 23, 119, 121
Scotland camp: 28
Screen ape: 197
Seeley, Henry: 166
Self-dumping skip: 84
Seneca, Missouri: 7, 28, 38
Sergeant, J. B.: 31, 35
Shaft: 75, 84
Shafting: 83, 93
Shaft pumping system: 86
Shaft sinking: 73, 84
Shaner, Dolph: 20, 32, 34, 198
Shawnee Indians: 38
Sheet ground: 11, 45, 78
Sheet-ground mines: 83, 88
Sherwood Diggings: 28, 165
Shoal Creek: 3, 18, 19, 80, 115, 122
Shoal Creek mines: 23
Shoemaker, Lloyd C.: 15
Short Creek: 104
Short Creek valley: 11
Shoveler: 71, 76, 89, 92, 98, 200
Siebenthal, C. E.: 6
Silicosis: 72, 93, 182, 185, 188, 189, 190, 194, 195
Six Boils country: 18
Skip pocket: 94
Slabs: 88
Slag: 119, 126
Sluice boxes: 264
Sluicing: 101
Slusher-loader: 94, 95
Smelter: 30, 32, 104, 108, 113, 121, 160, 161, 175, 192
Smelterman: 76, 229
Smelting: 23, 41, 100, 113, 117, 121, 122, 166
Smith, W. S. T.: 5
Smithfield, Missouri: 28, 37
Social conditions: 249
Social problems: 190, 194
Social Security: 191
Socialist party: 230
Soft ground: 11
Southside Tract: 166
Southwest district of Missouri: 4
Southwestern mining camps: 24
Southwestern Missouri Safety and Sanitation Association: 183
Southwestern Power Company: 80
Southwest Missouri and Southeast Kansas Lead and Zinc Miners' Association: 175, 261
Southwest Missouri Electrical Railway: 83

Southwest Missouri Safety and Sanitation Association: 185
Spearing, Dr. Joseph: 194
Spring City, Missouri: 28
Spring River: 3, 18, 24, 81
Spurgeon, Missouri: 50, 115, 209
Stark City, Missouri: 28
Star rig: 43
Steam engine: 43, 70, 83
Steam hoisting equipment: 70, 130
Steam power: 67
Steam pumps: 74
Steam shovels: 89
Stephens Diggings: 28, 165
Sterling Company: 150
Stockholders: 176
Stoddard, Benjamin: 16
Stone, Irene G.: 128–29
Stope method: 87–93
Strike: 232, 234, 245, 252
Strike-breaking: 231
Strip-pit operation: 96, 172
Sucker Flats camp: 96
Sultana Mining Company: 168
Supports: 88
Swallow, George C.: 22, 78, 123
Swedish Syndicate: 148
Swindle Hill camp: 28, 37

Tabling: 108
Tailing piles: 82, 108, 110
Taney County, Missouri: 80
Tanyard Hollow diggings: 20, 165
Tar Creek: 7
Tar River camp: 28
Tariff protection: 178, 270
Tarr, William A.: 6
Taylor, F. A.: 92
Taylor Diggings: 28, 165
Teamsters: 85
Thompson and Graves Mining and Royalty Company: 147
Thoms Station camp: 28, 82, 96
Thurman Diggings: 28, 37, 147
Thousand Acre Tract camp: 37, 150, 255
Tiffin, Edward: 16
Time study: 92
Tingle, William: 19, 116
Titus, Cliff: 204
Tonnage: 92
Tramming: 90
Transportation: 24, 82
Transport facilities: 111
Trappers: 113
Treece, Kansas: 28, 95, 195
Tripod drills: 87
Tri-State camps: 27
Tri-State Metal, Mine and Smelters Union: 234
Tri-State Milling: 100
Tri-State Mine Finance: 160
Tri-State Miners: 29
Tri-State Ore Producers' Association: 40, 172, 177, 178, 182, 187, 192–93, 194, 232, 233
Tri-State prospecting: 41
Tri-State smelting: 113
Tri-State society: 249
Tri-State survey committee: 250
Tri-State Zinc Company: 243
Trolley lines: 82, 83
Tuberculosis: 182, 185, 194, 195
Tub hooker: 90, 197
Tubs: 88
Turkey Creek: 3, 18, 30, 37, 85, 101, 117
Turkey Creek camp: 20, 24, 26
Twin Grove Community: 117
"Two-man power mining": 127

Underground lighting: 92
Underwriters Mining Company: 170
Union Army: 25, 26
Union City, Missouri: 35
Unionization: 196, 201, 210, 248, 265
United Cement, Lime, and Gypsum Workers (AFL): 246

United States Bureau of Indian Affairs: 128, 154
United States Bureau of Labor Statistics: 244
United States Bureau of Mines: 93, 97, 155, 162, 169, 177, 187, 250
United States Department of Labor: 190
United States secretary of the interior: 154
United States Mining and Smelting Company: 150
United States Steel Corporation: 174
United Steel Workers of America (CIO): 40, 247
United Zinc Company: 167

Van Buren, Arkansas: 171
Van Hise, C. R.: 5, 41, 79, 162
Ventilation: 70, 180
Victor Mining Company: 153
Vinegar Hill Mining Company: 170
Vivian and Sons, Wales: 174–75

Waco, Missouri: 28
Wages: 181, 199, 200
Walking Beam Pump: 86
Wallower Mining Company: 170
Warren, G. Ed: 240
Water deposition theory: 5
Waters, Sylvester: 236
Webb City, Missouri: 28, 30, 37, 74, 78, 83, 96, 106, 148, 153, 165, 175, 184, 192, 204, 258
Webb City–Carterville–Porto Rico–Duenweg area: 12
Webb City–Duenweg–Oronogo district: 209
Webb City Mining Company: 167
Webb City–Oronogo area: 172
Weidman, Samuel: 6
Wentworth, Missouri: 3, 28
Western Federation of Miners: 227, 228
Western Journal: 24
Western Journal and Civilian: 24, 162
West Joplin Lead and Zinc Company: 165
Whip hoister: 69
Whisky: 251
White lead: 120
White River power plant: 81
Wildcat prospect: 40
Wildcat shaft: 91
Wilfrey tables: 108
Williams, Walter: 15, 202, 252
Windlass: 69, 74
Winslow, Arthur: 21, 102
Witcombe, Wallace H.: 130, 161
Worker efficiency: 92, 182
Worker hazards: 181
Worthington Steam Pump: 86
Wright, Clarence A.: 5, 79, 151, 162, 169, 181, 202

Yellow Dog Mining Company: 106

Zinc-Lead Company of America: 166
Zinc refining: 126
Zinc smelter: 124, 171, 174
Zook, Jesse: 81, 176, 202